Alternative
SCIENCE

Alternative SCIENCE

Challenging the Myths of the Scientific Establishment

▼

RICHARD MILTON

Park Street Press
Rochester, Vermont

Park Street Press
One Park Street
Rochester, Vermont 05767
www.InnerTraditions.com

LIBRARY OF CONGRESS CATALOGING-IN-PUBLICATION DATA
Milton, Richard, 1943–
    [Forbidden science. English]
    Alternative science : challenging the myths of the scientific
establishment / Richard Milton.
       p. cm.
    Originally published: Forbidden science : Fourth Estate Ltd, 1994.
    Includes bibliographical references and index.
    ISBN 0-89281-631-7 (alk. paper)
    1. Science—Philosophy. 2. Fundamentalism. 3. Toleration. I. Title.
    Q175.M632513 1996
    501—dc20   96-4983
                                                      CIP

Printed and bound in the United States

10 9 8 7 6 5 4 3 2

Park Street Press is a division of Inner Traditions International

Distributed to the book trade in Canada by Publishers Group West (PGW),
    Toronto, Ontario

Contents

Preface

When *Alternative Science* was first published in Britain in April 1994, some critics thought that its central idea was too extreme to be taken seriously. I proposed that we are living through an era of increased academic intolerance in which a scientific fundamentalism has infected many American and British universities, a malady as virulent and pernicious in its way as today's tide of religious fundamentalism.

Yet within weeks of publication, an event took place on British and American television that confirmed my worst fears and showed that my thesis was not too far fetched, but all too horrifyingly real.

Television networks on both sides of the Atlantic broadcast an hour-long science special entitled *Too Close to the Sun,* a film that examined the intense controversy surrounding the discovery by professor Martin Fleischman and professor Stanley Pons of the much disputed phenomenon of "Cold Fusion"—a controversy I examine in detail in Chapter Three of this book.

The film was admirably balanced and included interviews with experts on both sides of the question. Halfway through, the film showed an interview subject who is a distinguished senior American physicist from an equally distinguished American research institution. There is nothing unexpected or controversial about such an appearance. But in this case, because of the subject matter, the scientist chose to appear in silhouette, his identity disguised.

Remember this was not *60 Minutes* but a science programme. He was no Cosa Nostra bag man or terrorist informant on the run. He was a professional scientist who was concerned that if his identity was publicly revealed, and if the institution that employed him discovered that he had been spending research funds on a forbidden subject like cold fusion, then his funding, and even his tenure, might be in jeopardy.

Sadly, his fears have been fully justified by recent events. Dr Robert Jahn was demoted as Dean of Engineering at Princeton because his engineering laboratory discovered repeatable empirical evidence of the effect of human consciousness on electronic instruments. Dr Jacques Bienveniste was dismissed from his post with France's prestigious Pasteur Institute because his team discovered repeatable empirical evidence to support the fundamental principle of homeopathy. Dr John O'Bockris, professor of chemistry at Texas A&M University, was threatened with dismissal merely for researching the subject of cold fusion.

The scientific fundamentalism of which these are disturbing signs is found today not merely in remote provincial pockets of conservatism, but at the very top of the mainstream management of science on both sides of the Atlantic, as examples in later chapters show.

Every age of license is followed by an age of authoritarianism. The open, experimental, libertarian decades of the 1960s and 1970s have been succeeded by the cynical 1980s and 1990s. During these cycles of changing social attitudes academic liberty—freedom of thought—ebbs and flows on a tide of intolerance.

In times such as these, some academics appoint themselves vigilantes to guard the gates of science against troublemakers with new ideas and new discoveries. Yet science has a two-thousand-year record of success not because it has been guarded by an Inquisition but because it is self-regulating. It has succeeded because bad science is driven out by good: an ounce of open-minded experiment is worth any amount of authoritative opinion by self-styled scientific rationalists.

We are again living in a time of rising academic intolerance in which important new discoveries in physics, medicine, and biology are being ridiculed and rejected for reasons that are not scientific. Once again, something precious and irreplaceable is under attack.

PART ONE

....

Science

Perfectly exact physics is not so very exact
just as holy men are not so very holy

WILHELM REICH

CHAPTER ONE

Too Wonderful to be True

Nothing is too wonderful to be true,
if it be consistent with the laws of nature.
MICHAEL FARADAY

On the night of 26 April 1803 the people of the rural French village of L'Aigle were turned out of their beds by the thunderous noise of more than two thousand rocks falling from the sky. This phenomenal shower of meteorites also woke up the members of the Académie Française. They appointed a commission to investigate the incident and, when its report was published, the French scientists were compelled to admit that stones really could fall from the sky.

Throughout the previous century, the enlightenment of the scientific age had dispelled, one after another, the medieval illusions and old wives' tales fostered by generations of country people and town dwellers alike. In the cold light of modern post-Newtonian science, the idea that stones could fall out of the sky had been denounced as an unscientific absurdity by the Académie, Europe's leading rational authority. Antoine Lavoisier, father of modern chemistry, told his fellow Academicians, 'Stones cannot fall from the sky, because there are no stones in the sky!'

Museums all over Europe had thrown out their cherished meteorite specimens with the rubbish as humiliating reminders of a superstitious past. Today scarcely a single specimen is known that predates 1790, except for the 280-pound stone that fell in Alsace in 1492, that is kept in the town hall of Ensisheim, and that proved too heavy for even the Académie Française to dislodge.

Despite final acceptance of the phenomenon by French scientists following the L'Aigle shower, acceptance elsewhere was slow and reluctant. The American president, Thomas Jefferson (who had studied natural sciences), refused to believe in the extraterrestrial origin of a meteorite that fell in Weston, Connecticut, as late as 1807.

One of the first professional scientists to take meteorites seriously was the German chemist and metallurgist Karl Reichenbach. Reichenbach made quite a name for himself in Germany's embryonic chemical industry. He was the first chemist to isolate important materials by distilling coal tar, the method by which he discovered creosote and paraffin, the principal lighting oil of the nineteenth century. He also studied the metallurgy of meteorites.

Having broken new ground in several areas of chemistry and having demonstrated the value of empirical research in dispelling superstition (even scientifically approved superstition) Reichenbach struck out in 1844 into the territory that was to prove his undoing. He was introduced by a doctor in Vienna to a sick young woman who appeared to be able to perceive the field surrounding a magnet. He carried out a number of blind experiments with her to test this ability and confirmed that it appeared to be real. He published his findings in 1845 – his book becoming a sensation throughout European society.[1]

European science, on the other hand, took a distinctly dim view of this research. Stones from the sky was quite enough originality for one man: a girl who could see magnetic fields was going too far. Reichenbach retreated back to more orthodox research in physical chemistry and disappeared into obscurity. Today his name is rarely to be found in any reference book or encyclopaedia. You can read as much as you wish about the inventor of the frozen pea or the discoverer of purple dye, but you will find scarcely a word about the scientist who discovered paraffin.

Karl Reichenbach represents all that is splendid about scientific endeavour and the courageous investigation of the new and the unknown. At the same time he symbolises the extraordinary nature of the resistance that we all feel to the new. It was science that proved the reality of meteorites; yet it was science that had for a generation dismissed them as fanciful. How can the same principles, the same methods – even the same people – reach diametrically opposed conclusions?

When the nineteenth century's greatest experimental physicist, Michael Faraday, announced that he had found a new source of energy

4

simply by moving a magnet in a coil of wire, many educated people found the claim impossible to believe and looked on the young man as a charlatan. Faraday responded with perhaps the most memorable words ever uttered by a scientist: 'Nothing is too wonderful to be true, if it be consistent with the laws of nature.'

Only twenty years after Faraday made the breakthrough in electromagnetism that made the modern world possible, the Fox sisters of New York began to hear rapping at their table from 'departed spirits' – the first modern recorded instance of paranormal phenomena. Faraday left his researches and spent substantial time and energy attempting (without success) to prove the Fox sisters fraudulent. Even someone who professed to believe that nothing is too wonderful to be true found table-turning an intolerable affront to reason.

It seems that when it comes to investigating natural phenomena there is a line that some scientists, for some reason, are unwilling to cross. Equally, it seems that there are some individuals, including very distinguished scientists, who are willing to risk the censure and ridicule of their colleagues by stepping over that mark. This book is about those scientists. But, more importantly, it is about the curious social and intellectual forces that seek to prohibit such research; about those areas of scientific research that are taboo subjects: subjects whose discussion is forbidden under pain of ridicule and ostracism.

It is also about what I believe to be a worrying but well-documented social trend; a trend towards a normalised world view based on a singular model that is derived entirely from the reductionist western scientific viewpoint, and the marginalisation and suppression of any form of scientific dissent or alternative world view.

From the examples given earlier you might imagine that I am speaking historically and that, while the ill-informed people of previous centuries fell into the error of rejecting major discoveries from the worlds of electricity and astronomy, no scientist today would react in such an intemperate, unreflecting way about a matter that must be purely a question of fact. Actually, Faraday and Reichenbach would almost certainly have experienced *more* difficulty not less in making their voices heard in today's climate of intolerance.

In March 1989, Professor Martin Fleischmann of Southampton University and Professor Stanley Pons of the University of Utah put a new phrase into the scientific lexicon when they jointly announced the discovery of 'cold fusion' – the production of usable amounts of energy by what seemed to be a nuclear process occurring in a jar of water at room temperature.[2]

The reaction to the announcement was almost universally hostile. The two were ridiculed by both the popular and the scientific press, especially *Nature* magazine. Major institutions who had already spent several billion dollars in pursuit of 'hot' fusion – notably Harwell and MIT – announced that Fleischmann and Pons's results could not be reproduced. When it was discovered that MIT had fudged their experimental results (as described in Chapter 3) they merely amended their conclusion from 'failure to reproduce' to 'too insensitive to confirm'. But by that time the damage had been done: cold fusion had been discredited. No more significant research money was to be granted for cold fusion research and the United States patent office still relies on the MIT findings to reject all patent applications involving cold fusion.

This official position remains despite the fact that, at the time of writing, cold fusion reactions have been reproduced by ninety-two major universities and commercial corporations in ten countries around the world including Stanford Research Institute, Los Alamos National Laboratory, Oak Ridge National Laboratory, Naval Research Laboratory, Naval Weapons Centre at China Lake, Naval Ocean Systems Centre, Texas A & M University, California Polytechnic Institute and Japan's Hokkaido and Osaka Universities.[3]

In many ways cold fusion is the perfect paradigm of scientific taboo in action. The high priests of hot fusion were quick to ostracise and ridicule those whom they saw as profaning the sacred wisdom. And empirical fact counted for nothing in the face of their concerted derision.

Even more disturbing are cases from the world of medical research. Respected journals like the *Lancet* and *British Medical Journal* have published scores of studies reporting evidence for the effectiveness of holistic methods of prevention and treatment of cancer, heart disease and other illnesses; yet these studies have largely been ignored by those funding the fight against such diseases, because they seem to dabble in another taboo area: the idea of a causal link between mind and body.[4]

Often those who cry taboo do so from the best of motives: a desire to ensure that our hard-won scientific enlightenment is not corrupted by the credulous acceptance of crank ideas and that the community does not slide back into what Sir Karl Popper graphically called the 'tyranny of opinion'.[5] Yet in setting out to guard the frontiers of knowledge, some scientific purists are adopting a brand of skepticism that is indistinguishable from the tyranny they seek to resist. These modern skeptics are sometimes the most unreflecting of individuals yet their devotion to the cause of science impels them to appoint themselves guardians of the spirit of truth. The nature of these guardians and their activities is examined in Chapter 10.

Equally, of course, those who rightly value the importance of knowledge are concerned at the damage that crank beliefs can do to any community and are right to take action to call those with revolutionary ideas to account, using the principles of the scientific method and its rigorous demand for empirical evidence. And this raises the important question of just how we can tell a real crank from a real innovator – a Faraday from a false prophet – an issue that is explored in Chapter 11.

The cases of scientific taboo referred to earlier – and many others described in later chapters – raise a number of important questions of general public interest. Who do you have to be to have a voice about scientific research on which large sums of public money are spent? Who *decides* who you have to be? In what forum, or by what mechanism, can the voices of scientific dissent ever be heard in Britain today?

In past centuries these questions were less vital because the arguments were about whether the Earth was round or flat – interesting but not life-threatening. Now the arguments are about whether HIV causes AIDS, whether we are heading for a global warming catastrophe, and whether disease can be treated holistically. Even a question such as whether cold fusion is real or illusory has a hard political edge and hefty financial implications for the taxpayers who foot the research bill.

This year Britain will spend more than £5,000 million of public money on scientific research, almost 2 per cent of the nation's overall budget, more than twice what will be spent on Wales and equivalent to the amount spent on Scotland.[6]

These billions will be spent on a huge variety of projects, some in pure science whose aim is simply the advancement of knowledge and whose ultimate benefits it is impossible to predict. But much of this money will be spent on research aimed at relatively short-term benefits to the community here and now, such as medical research. In the past twenty years Britain has spent £2,000 million on medical research (at current prices). Yet in that twenty years, and regardless of such expenditure, *there has been no change in life expectancy*, despite often repeated claims to the contrary. Even more disturbing, in the same period there has been *no significant reduction* in deaths due to the most common forms of cancer or heart disease. Indeed, studies in Britain and the United States show that deaths through cancer have increased over the past thirty years (as described in Chapter 7) again counter to claims that medical research is improving treatment.[7]

Later chapters tell how, in part, the failures of medical science can be attributed to an unwillingness to consider and act on the mountains of evidence showing that a substantial part of our research effort could more profitably be redirected to prevention rather than cure, and that holistic methods of treatment are effective.

Of course, not all members of the scientific establishment are deaf to new discoveries – far from it. Many scientists are themselves in the forefront of the struggle to topple the barricades that some of their colleagues have erected. But in some ways big science, institutional science, is gaining many of the trappings of a banana republic dictatorship: a revenue of billions that is unaccounted for and an administration that is unaccountable to tax payers, except in a cosmetic way; the making and unmaking of reputations by a tame scientific press; the scientific police who make sure members of the profession are thinking along politically correct lines, and who patrol the content of scientific publications.

Press and broadcasters have not hesitated to ask whether public expenditure on the Royal Family is justified and whether we, the community, receive value for the £10 million or so we pay each year for the privy purse. It seems to me that there is a far more important question going entirely unasked: are we, the community, receiving value for the £5,000 *million* of public money spent each year on scientific research?

The truly disturbing thing about this question is that, as things stand, anyone who is not a professional scientist (and precious few who are) simply has no way to answer it. And unless and until we have some mechanism by which the community can call institutional scientists to account, not just administratively, but in terms of the value of their research, it will remain impossible for us to know whether our money is being spent wisely or foolishly.

At present the only mechanism that exists for overseeing scientific research and the allocation of huge sums of public money to pay for such research is the system of peer review, under which papers reporting results and proposing future research are reviewed by their authors' scientific colleagues before being accepted for publication and before the allocation of further resources. The system has come about because, on the face of it, it makes sense as the most efficient method.

The peer review consensus seemed appropriate to a gentleman's profession. After all, surely only a nuclear physicist is suitably qualified to decide whether public money should be spent on researching cold fusion? Only a cancer specialist should decide whether medical research funds should be spent on holistic treatment? Isn't it appropriate that scientists should regulate science in the same way that lawyers oversee the legal profession, and journalists supervise the press? The events described in detail later show that, however precise their calculations may be in the laboratory, scientists' arithmetic is no better than anyone else's when it comes to holding the purse strings. And while it can be argued that lawyers and journalists are no better at regulating themselves, they, unlike scientists, are not responsible for spending thousands of millions of pounds of public money each year – they spend their own money, not ours.

We live in a society that prizes liberty of conscience and freedom of speech. Plurality of viewpoint is now so highly valued it has even been enshrined in the legislation governing British television broadcasting, the act of 1990 obliging broadcasters to provide a 'broad range and diversity of independent productions'. It is an article of faith in every department of public life that diversity of opinion, tolerance and a robust openness are the vital signs of our society's health: that no subject for debate should be forbidden and that no person or group should have the authority to deny such discussion.

Yet one great taboo area remains; one area of public life that may be spoken of only by its high priests and whose inner sanctum the rest of us may not approach; the area of institutionalised science – big science.

This book examines some of the most important cases of scientific discoveries being treated as taboo subjects; looks at the cost of this ideological 'correctness' to the community and – most ambitious of all – attempts to identify the causes of this behaviour by both scientists and non-scientists. It is an attempt by one outsider to pay a profane visit to that inner sanctum on behalf of the millions who will never get a chance to see for themselves what they are paying for.

I want to begin by examining in detail some well documented cases of the taboo reaction in science and their effects on the community. But first, let's take a quick tour of the subject by looking briefly at some classic cases of scientific taboo in action.

....

CHAPTER TWO

A Completely Idiotic Idea

Edison's electric lamp is a completely idiotic idea.
SIR WILLIAM PREECE FRS
Post Office Chief Engineer

For five years, from December 1903 to September 1908, two young bicycle mechanics from Ohio repeatedly claimed to have built a heavier-than-air flying machine and to have flown it successfully. But despite scores of public demonstrations, affidavits from local dignitaries and photographs of themselves flying, the claims of Wilbur and Orville Wright were derided and dismissed as a hoax by the *Scientific American*, the *New York Herald*, the US Army and most American scientists.

Experts were so convinced, on purely scientific grounds, that powered heavier-than-air flight was impossible that they rejected the Wright brothers' claims without troubling to examine the evidence. It was not until President Theodore Roosevelt ordered public trials at Fort Myers in 1908 that the Wrights were able to prove conclusively their claim and the Army and scientific press were compelled to accept that their flying machine was a reality.[1]

In one of those delightful quirks of fate that somehow haunt the history of science, only weeks before the Wrights first flew at Kitty Hawk, North Carolina, the professor of mathematics and astronomy at Johns Hopkins University, Simon Newcomb, had published an article in *The Independent* which showed scientifically that powered human flight was 'utterly impossible.'[2] Powered flight, Newcomb believed, would require the discovery of some new unsuspected force in nature. Only a year earlier, Rear-Admiral George Melville, chief engineer of the US Navy, wrote in the *North American Review* that attempting to fly was 'absurd'. It was armed with such eminent testimonies as these that *Scientific American* and the *New York Herald* scoffed at the Wrights as a pair of hoaxers.

It is not so surprising that a couple of young bicycle mechanics in a remote mid-western prairie town should be ignored by east coast city publishers at a time when the horse was still the principal means of transport. What is surprising is that the local newspapers in their home town of Dayton, Ohio, should have steadfastly ignored the Wrights. In 1904, Dayton bank president Torrence Huffman allowed the brothers to use a large tract of farm land owned by him outside the town for their flying experiments. The land was bordered by two main highways and the local railway line so that, as the months went by, hundreds of people actually saw the Wrights airborne in their flying machine. Two such spectators in 1905 were the general manager of the railway line, together with his chief engineer, who happened to be travelling past the field when Orville was airborne and ordered the conductor to stop the train so that they – and all the other passengers – could watch the machine in flight.

Many of these bewildered witnesses visited or wrote to the local newspapers to ask who were the young men that were regularly flying over 'Huffman Prairie' and why nothing had appeared about them. Eventually the enquiries became so frequent that the papers complained of their becoming a nuisance, but still their editors showed no interest in the story, sending neither a reporter nor photographer.

In 1940, Dan Kumler, the city editor of the Dayton *Daily News* at the time of the flights, gave an interview about his refusal to publish anything thirty-five years earlier and spoke frankly about his reasons. Kumler recalled, 'We just didn't believe it. Of course, you remember that the Wrights at that time were terribly secretive.'

The interviewer responded incredulously, 'You mean they were secretive about the fact that they were flying over an open field?' Kumler considered the question, grinned, and said, 'I guess the truth is we were just plain dumb.'[3]

Kumler was far from alone in being dumb. The owner of the Dayton *Daily News*, James Cox, also admitted that, 'none of us believed the reports'. The managing editor of the Dayton *Journal*, Luther Beard, checked the reports personally with Orville Wright who told the newspaperman that he had flown for nearly five minutes that day. But Beard decided the story must be a mare's nest and his paper, too, carried no report.

In January 1906, more than two years after the Wrights had first flown, *Scientific American* carried an article ridiculing the 'alleged' flights that the Wrights claimed to have made. Without a trace of irony, the magazine gave as its main reason for not believing the Wrights the fact that the American press had failed to write anything about them.

> If such sensational and tremendously important experiments are being conducted in a not very remote part of the country, on a subject in which almost everybody feels the most profound interest, is it possible to believe that the enterprising American reporter, who, it is well known, comes down the chimney when the door is locked in his face – even if he has to scale a fifteen-story skyscraper to do so – would not have ascertained all about them and published them broadcast long ago?[4]

Although not willing himself to climb down the chimney in pursuit of the scientific scoop of the century, the magazine's editor – as an afterthought to this editorial – adopted the simpler and less sooty expedient of writing a letter to the Wrights asking if the reports were perhaps true after all.

One way of explaining this odd reluctance to come to terms with the new, even when there is plenty of concrete evidence available, is to appeal to the natural human tendency not to believe things that sound impossible unless we see them with our own eyes – a healthy skepticism. But there is a good deal more to this phenomenon than healthy skepticism. It is a refusal even to open our eyes to examine the evidence that is plainly on view. And it is a phenomenon that occurs so regularly in the history of science and technology as to be almost an integral part of the process.

On 4 July 1897, Britain's Grand Fleet, the world's most powerful battle fleet, was assembled at Spithead to be reviewed by Prince Edward the Prince of Wales, representing Her Majesty Queen Victoria on the occasion of her diamond jubilee. Anchored in five rows, each five miles long, were 166 capital ships of the line manned by 50,000 sailors and marines, each warship flying congratulatory signals to the sovereign. The hundreds of other craft at the Spithead review – both naval and civilian vessels – were kept in strict order by patrolling naval torpedo boats; the fastest vessels afloat.

Suddenly, to the astonishment of the Prince, the naval brass and the thousands of sightseers, a small boat, only 100 feet long, broke away from her position on the sidelines and entered the long sea lane between the carefully measured ranks of battleships. Two torpedo boats were dispatched at once by an indignant admiral to run down the interloper but to the amazement of the royal party the little craft, bearing the unfamiliar name *Turbinia* on her stern, began to pull rapidly away from her pursuers. The Navy vessels squeezed out every ounce of steam from their boilers but the little *Turbinia* accelerated impertinently away, putting on a show for the appreciative crowds.[5]

The tiny craft easily outdistanced her pursuers, reaching the unheard-of speed of some 34 knots – just a little under 40 miles per hour and almost twice the speed of her contemporaries. At the helm of *Turbinia* as she tormented her naval pursuers was a marine engineer called Charles Parsons, whose claims had been ignored by the Admiralty and who had at last proved a long-cherished theory and made his unique vessel famous before the whole world.

Parsons's triumph at Spithead followed years of scorn for his engineering ideas. Despite being of aristocratic birth – youngest son of the Earl of Rosse – Charles Parsons had no easy ride. At Cambridge, as an undergraduate in the 1870s, he dodged his formal lessons and instead constructed a cardboard model of the turbine he hoped to build. He announced to his skeptical fellow undergraduates that he planned to build an engine that would run twenty times faster than any other form of propulsion.

Working in industry after graduating, Parsons contrived to keep developing his idea for a turbine until he finally constructed a practical engine in 1884. He built *Turbinia* as an experimental vessel and by a process of trial and error found the most efficient way of transferring the turbine's motive power to a propeller shaft. This gave him the fastest vessel afloat and, in principle, gave Britain's naval and merchant vessels a priceless advantage over every other nation.

Parsons at once offered to place his invention at the disposal of the Admiralty, explaining that it could propel even the largest battleship at speeds of more than 20 knots. The Lords of the Admiralty thanked Parsons politely but explained that such speeds were impossible and that Parsons' turbine-powered vessel would be 'uncontrollable'. After more fruitless correspondence along these lines, Parsons decided that only a public demonstration could make the admirals realise their mistake, and in a moment of inspiration he selected the Spithead naval review as his opportunity. After his dramatic demonstration his invention was, of course, taken up by every navy and shipping line in the world. By 1905 turbine-powered ships had reduced the Atlantic crossing to Canada to five and a half days, and by 1910 the turbine-powered Cunarder *Mauritania* had captured the Blue Riband for the Atlantic crossing from Southampton to New York.

Their lordships at the Admiralty have become something of a byword for negligently rejecting key inventions offered to them on a plate, a reputation that in the nineteenth century they certainly merited. In 1813, Francis Ronalds presented them with the world's first electric telegraph to replace manual semaphore signalling stations. The secretary to the Admiralty, John Barrow, replied on behalf of the First Sea Lord that: 'Mr Barrow presents his compliments to Mr Ronalds and acquaints him . . . that telegraphs of any kind are now wholly unnecessary; and that no other than the one now in use, will be adopted.'

In 1864, industrialist Henry Bessemer pointed out to their lordships that the Russian Navy was now using steel shells that could easily penetrate the hulls of British naval vessels that were clad only in iron. The Navy did not react to this news for a further twelve years, during which time the entire British fleet could have been sunk in minutes by a handful of modern battleships.[6]

In 1900, the managing director of the Linotype company, Arthur Pollen, innocently stumbled into probably the greatest scandal of all. Pollen was on holiday in Malta where warships of the British fleet were conducting gunnery exercises, and he contrived to get himself invited on board the cruiser *Dido* as an observer. What he saw in those trials shocked him deeply. The Navy's biggest guns at the time were of 13.5 inch calibre. Guns of this size are capable of projecting a shell more than 15 miles, but in the Malta trials Pollen saw such guns employed by *Empress of India* that were quite unable to hit anything at a range of less than just 1,500 yards.

The reason, he quickly learned, was that these huge guns were being aimed by human hand and eye alone using rangefinding techniques little different from those employed by naval gunners in Nelson's day – there was no scientific method of laying a gun and predicting the fall of shells if the target was moving or if the gun platform itself was under way, as it was bound to be in any real naval engagement.

He at once wrote to the Admiralty offering to design and manufacture a set of instruments that would make it possible to find the course and speed of any ship that could be kept under observation, up to a range of 20,000 yards (10 sea miles). He calculated that using such instruments, a naval commander would have more chance of hitting an enemy vessel at 10,000 yards than he currently had of hitting such a ship at 5,000 yards using manual methods.

Pollen was informed that his proposals were 'of no interest to their lordships'. Their reasons were not altogether without foundation, since the naval tactical thinking of the day favoured fighting, as in Nelson's time, at close quarters and little or no thought was given to long-range gunnery, especially as the Germans showed no interest in the subject. If they thought about it at all, their lordships were anticipating a gentlemanly scrap in which rival fleets would shell each other from within hailing distance.

Pollen continued to press the matter with the Admiralty and continued to be told to mind his own business. In 1905, when the Japanese fleet annihilated the Russian fleet by gunfire at Tsushima at a range of three miles, Pollen felt that the Admiralty might have changed its view and wrote again. This time he was told that the success of the Japanese was the result of their *not* having any system:

> Had there been a well organised system by which all guns fired together any mistake made by the system would have been repeated by every gun. There being no system there were a great number of misses but so many shots were fired and with such a fine freedom of selection that there was inevitably a good proportion of hits.[7]

It was thus the *absence* of a fire control system that had won the battle for the Japanese according to the Admiralty! Pollen wrote back to say that the Admiralty's approach to gunnery reminded him of Charles Lamb's story of the Chinese burning down their houses in the hope that a pig would be roasted in the process.

Pollen went on, at his own expense, to develop the world's first (mechanically) computerised system of naval fire control, but it was to be 1925 before capital ships of the navy were equipped with his system. In the meantime, the increasing range and explosive power of the torpedo meant that warships had to engage each other at longer and longer range: 14,000 to 16,000 and even 18,000 yards. The British fleet sailed into action at Jutland in 1915 unequipped for a long-range gunnery battle and this, combined with its light deck armour, resulted in its suffering huge damage by comparison with the German fleet, damage which some naval historians think amounted to a tactical defeat.[8]

You might imagine that the frosty reception met with by people like Pollen, Parsons and the Wrights came about because they were unknown amateurs. But similar examples can be found in the cases of even the most famous professional inventors.

In 1879, Thomas Edison was riding the crest of a wave of newspaper publicity and public adulation as the 'Napoleon of Science' and the 'Wizard of Menlo Park'. By this date he had patented more than 150 inventions. Among them were the stock exchange ticker-tape machine, the quadruplex telegraph and the phonograph, which he unveiled in April 1878 and which had quickly captured the public imagination after he had demonstrated recorded sound at the White House and the offices of *Scientific American*.

Edison was now seeking new scientific worlds to conquer at his Menlo Park research laboratories in New Jersey, the world's first purpose designed R & D facility. He chose a problem that had already defeated many of the world's most eminent electrical engineers: the incandescent electric light. Carbon arc lights and other experimental forms of lamp had been in existence for twenty years, but none of them was practical or economical for domestic or public use, and lighting remained confined to paraffin lamps and gaslight. Edison resolved both to find a practical lamp and to find a way of lighting them in parallel, so that if one lamp burnt out or was switched off, the circuit was not broken; a street or a house would not be plunged into darkness by the failure of one bulb.

Edison devoted most of 1879 to the problem. He and his team of research assistants tried an open-air filament, a vacuum and a partial vacuum, but none of them worked satisfactorily. He tested a huge variety of materials as filaments before eventually trying a carbonised thread in a vacuum, which to his immense joy burned continuously for forty hours. The secret that Edison had discovered was the use of relatively high resistance filaments (around 100 ohms as distinct from the 5 ohms or so of the carbon arc) to achieve his parallel circuit.

When the 'Napoleon of Science' announced to the world that he had succeeded in making a practical incandescent lamp, the reaction from professional scientists was one of stark disbelief, partly because they could not accept that such a high resistance filament would work. England's most distinguished electrical engineer Sir William Siemens, who had been working on electric lighting for some ten years, said that: 'Such startling announcements as these should be deprecated as being unworthy of science and mischievous to its true progress.'[9]

Once he had perfected his lamps, Edison rigged up a public demonstration by lighting the streets of Menlo Park around his laboratories where, for the first time in history, the night was ablaze with electricity. Members of the public travelled for miles to see the sight, but no scientist took up Edison's invitation to come and see his lamps in operation. Not even Professor Henry Morton, who lived nearby and was personally acquainted with Edison, bothered to travel the short distance to witness his lighting, but wrote that he felt compelled 'To protest in behalf of true science,' that Edison's experiments were 'a conspicuous failure, trumpeted as a wonderful success. A fraud upon the public.'

A Professor Du Moncel said, 'One must have lost all recollection of *American hoaxes* to accept such claims. The Sorcerer of Menlo Park appears not to be acquainted with the subtleties of the electrical science. Mr Edison takes us backwards.'

Edwin Weston, a respected specialist in arc lighting called Edison's claims, 'so manifestly absurd as to indicate a positive want of knowledge of the electric circuit and the principles governing the construction and operation of electrical machines.'

While the sightseers stood innocently basking in the radiance of Edison's luminous laboratory, Sir William Preece read a paper at the gloomy, gas-lit rooms of the Royal Society in London explaining that the kind of parallel lighting circuit that Edison was trying to perfect could never be technically feasible. Preece, the chief engineer of Britain's Post Office, who had studied under Faraday, left his distinguished audience in no doubt of his professional opinion. Edison's lamp and parallel circuit, he said, was 'a completely idiotic idea'. (Preece had also been chief engineer three years earlier when the Post Office rejected the telephone on the grounds that Britain had 'plenty of small boys to run messages'.)

Of course, the derision of his contemporaries failed to prevent Edison's invention from being universally adopted. This was largely because financiers such as J.P. Morgan and William Vanderbilt had enough respect for Edison to invest in his inventions whatever other scientists might say. But there are cases where equally respected inventors succumbed to such derision.

To the American public of the 1880s and 1890s, Nikola Tesla was as well known as a scientific wizard as Edison. Indeed, it was Tesla who made alternating current generation and distribution practical, and who thus allowed Westinghouse to amass a fortune in the early electrical industry (Edison unwisely backed direct current generation).

However, Tesla also unveiled a steam turbine at about the same time as Charles Parsons was finally getting his ideas successfully adopted in England. Unlike the Parsons design, which has influenced all subsequent turbine and jet engine design, Tesla's turbine did not use vanes to direct the flow of gases but used circular flat plates which relied on turbulence to make them rotate. There is still much debate among engineers as to the efficiency of Tesla's design, but some researchers believe that it is more efficient than the classic Parsons design, while undoubtedly being many times cheaper and simpler to construct and maintain. Yet despite his eminence as an inventor, Tesla was unable to secure any financial backing for his turbine, which remains merely a scientific historical curiosity.

Twentieth century inventors fared little better. In 1912, an inventor from Adelaide, Australia, E.L. de Mole, sent to the British War Office his plans for a tracked armoured vehicle capable of transporting soldiers – the tank. With rare vision, de Mole had foreseen that the key tactical problem in the next war would be how to cross no-man's land in the face of deadly automatic fire. His solution was, of course, the one finally adopted. The War Office – an organisation whose sole function was to prepare for a future war – rejected his offer. They rejected the offer again when he repeated it in 1915, by which time men were dying on the Western Front at the rate of 1,000 a day for lack of suitable armoured protection.[10]

A number of individuals were to play a part in the development of the tank and many of them were accorded some recognition by the Royal Commission on Awards to Inventors which sat in 1919. Prominent among them were Winston Churchill, then First Lord of the Admiralty, Lieutenant Colonel Swinton and Commodore Murray Sueter. Interestingly, there were just as many supporters of the idea in the Navy as in the Army – and just as many opponents. At first the inventions were regarded as landships, hence Churchill's interest, and the Navy's involvement. But Murray Sueter records the Fourth Lord of the Admiralty as saying in 1915:

> Caterpillar landships are idiotic and useless. Nobody has asked for them and nobody wants them. Those officers and men are wasting their time and are not pulling their proper weight in the war. If I had my way I would disband the whole lot of them. Anyhow I am going to do my best to see that it is done and stop all this armoured car and caterpillar nonsense.[11]

His lordship had counterparts in the Army in the form of the director of artillery and the assistant director of transport, both of whom staunchly resisted the tank and continued to believe in the power of horsed cavalry and the prolonged artillery barrage, although as early as 1915 there was overwhelming evidence for the ineffectiveness of both under modern conditions of war.

Did the spate of inventions stimulated by the First World War make orthodox science more receptive to innovation? Judging by events in the inter-war years it seems not to have had any such effect. One evening in January 1926, forty or so members of Britain's most august scientific body, the Royal Society, accompanied by newspaper reporters from *The Times* and the *Daily Chronicle*, found themselves trooping up a dingy narrow concrete staircase to a cramped second-floor room in Frith Street, in London's Soho. These distinguished visitors had come to the unaccustomed surroundings of the notorious red light district at the invitation of John Logie Baird, to witness a demonstration of his new 'televisor' apparatus.

In his Frith Street laboratory Baird had been working for some four years to develop a practical television system, and had now set up his first major public demonstration. To prevent industrial espionage, he had concealed his transmitting apparatus behind a curtain, so the assembled scientists were presented with merely a receiving set to watch. According to the reporter from *The Times*:

> First on a receiver in the same room as the transmitter and then on a portable receiver in another room, [appeared] the recognizable head of a person speaking. The image as transmitted was faint and often blurred but substantiated a claim that through the 'Televisor', as Mr Baird has named his apparatus, it is possible to reproduce instantly the details of movement and such things as the play of expression on the face.[12]

Apart from the acknowledgement that Baird had substantiated his claim, the newspaper made no comment on the invention. The assembled scientists, however, had a number of comments. Present at the demonstration was William Fox, a correspondent with the Press Association and friend of Baird who acted as his press adviser. Fox later recalled:

> I'd say they didn't believe it a bit. They thought it was all a trick or something equally disreputable. I did hear one fellow say Baird was a mere mountebank, merely after what he could get. Other comments were 'nothing much'; 'absolute swindler'; 'doesn't know what he's about'; and one fellow came out very definitely and said; 'Well, what's the good of it when you've got it? What useful purpose will it serve?'[13]

It is fair to add that the modern television system that we enjoy owes very little technically to Baird's electromechanical system which suffered from major limitations that would prevent high-definition transmissions. But even his stiffest critics accept that it was largely due to Baird's pioneering spirit and his energetic efforts that Britain introduced the world's first regular public television service in 1936. And I do not think that it was technical limitations that some members of the Royal Society had in mind when they described Baird as a swindler or asked what was the use of his invention.

Even innovators themselves can fall victim to the taboo reaction, suggesting that no one is entirely immune. Alan Campbell-Swinton, a key pioneer of the cathode ray tube for television, addressed the Radio Society of Great Britain in 1924 on the subject of 'Seeing at Distance'. As long ago as 1908, Campbell-Swinton had brilliantly foreshadowed electronic television in a letter to *Nature*. Yet only two years before Baird's public demonstration, he told his audience, 'It is probably scarcely worth anyone's while to pursue it.' And even the discoverer of radio waves, German physicist Heinrich Hertz, warned the young Guglielmo Marconi that he was 'wasting his time' experimenting with wireless broadcasting.

Professional scientists who pronounced powered flight impossible at the beginning of the century were still at it more than fifty years later. In September 1957, Britain's Astronomer Royal, Sir Harold Spencer Jones, was asked by a journalist what he thought the prospects were for space travel. Jones told him: 'Space travel is bunk.' Two weeks later, Sputnik 1 was launched into Earth orbit by the Russians.

Looking through the catalogue of prejudice and rejection represented by the examples above it is difficult to know whether to laugh or to cry. Somehow it is amusing to read of intelligent, well-informed people scorning such discoveries, perhaps because it makes our own prejudices easier to bear: American President Thomas Jefferson, the principal architect of the Declaration of Independence, who said, 'I would sooner believe that two Yankee professors lied, than that stones fell from the sky'; astronomer Simon Newcomb who wrote that flight was 'utterly impossible' weeks before the Wrights took to the air; Admiral William Leahy who told President Truman that 'The [atomic] bomb will never go off, and I speak as an expert in explosives.' All have earned a place in our hearts for being so completely, totally and gloriously wrong.[14]

Yet there is a serious side to this question; a side which may well be costing the community dear. What if Tesla's turbine really is more efficient and cheaper to manufacture and maintain than Parsons'?

Might not every electricity generating station in the world be cheaper to build and run? What might have happened at Jutland in 1915 if the Grand Fleet had been equipped with Pollen's computerised fire control system? Or the Somme in 1916 if the British Army had been well-equipped with de Mole's tanks? (And what effect would an allied victory have had on European history?) And, perhaps more importantly, what about the tens or hundreds or thousands of other discoveries that no one has ever heard about because they were derided into oblivion?

Anyone who thinks it impossible or unlikely that a scientific discovery of any real commercial significance could be foolishly consigned to the scrap heap merely through intolerance and misplaced skepticism should consider carefully the subject of the next chapter.

....

CHAPTER THREE

Sunbeams from Cucumbers

He had been eight years upon a project
for extracting sunbeams out of cucumbers.
JONATHAN SWIFT
Gulliver's Travels

No other scientific endeavour has consumed so much talent, so much cash and so many years of sustained effort as the race to harness the power that makes the Sun shine. Billions of pounds (and dollars, roubles and yen), more than four decades of research and the careers of thousands of physicists have been expended on the search for a nuclear reactor that will generate limitless power from the fusion of hydrogen atoms. There are grey-haired professors with lined faces still poring intently over the equations they first looked at eagerly with bright young eyes in the 1940s and 1950s. They will go into retirement with their dreams of cheap, safe power from fusion still years in the future, for the obstacles in their paths are as formidable now as ever.

Fusion is the process taking place in the Sun's core where, at temperatures of millions of degrees, hydrogen atoms are compressed together by elemental forces to form helium and a massive outpouring of energy in the thermonuclear reaction of the hydrogen bomb.

It is not difficult, then, to imagine how people who have invested their talent and their lives in the quest to tame such forces are likely to react when told that fusion is possible at room temperature, and in a jam jar.

Hydrogen atoms repel each other strongly – so strongly that no known chemical reaction can persuade them to fuse. There are, though, heavier isotopes of hydrogen, such as deuterium, which together with oxygen makes heavy water and which under the right circumstances can be made to fuse in nuclear reactions. When they do so, they release energy. However, the only circumstances so far under which hydrogen atoms have been persuaded to fuse have nothing in common with the measured calm of the laboratory bench but are more like a scene from Dante's *Inferno*. In the centre of the Sun and other stars, the atoms are squeezed by cataclysmic gravitational forces to form a plasma of the nuclei of hydrogen atoms at a temperature of millions of degrees. These high temperatures kindle a self-sustaining reaction in which hydrogen is 'burnt' as the fuel.

The scientific world was thus astonished when, in March 1989, Professor Martin Fleischmann of Southampton University and his former student, Professor Stanley Pons of the University of Utah, held a press conference at which they jointly announced the discovery of 'cold fusion' – the production of usable amounts of energy by what seemed to be a nuclear process occurring in a jar of water at room temperature.

Fleischmann and Pons told an incredulous press conference that they had passed an electric current through a pair of electrodes made of precious metals – one platinum, the other palladium – immersed in a glass jar of heavy water in which was dissolved some lithium salts. This very simple set-up (the *Daily Telegraph* later estimated its cost as around £90) was claimed to produce heat energy between four and ten times greater than the electrical energy they were putting in. No purely chemical reaction could produce a result of such magnitude so, said the scientists, it must be nuclear fusion. Further details would be revealed soon in a scientific paper.

Both scientists are distinguished in their field, that of electro-chemistry. But in making their press announcement they were breaking with the usual tradition of announcing major scientific discoveries of this sort. The usual process is one of submitting an article to *Nature* magazine which in turn would submit it to qualified referees. If the two chemists' scientific peers found the paper acceptable, *Nature* would publish it, they would be recognised as having priority in the discovery and – all being well – research cash would be forthcoming both to replicate their results and conduct further research.

But the two scientists perceived some difficulties. First, their paper would not be scrutinised by their exact peers because the discovery was unknown territory to electrochemists and indeed everyone else. It would probably be examined mainly by nuclear physicists – the men and women who had grown grey in the service of 'hot' fusion. This would be like asking Swift's 'Big Endians' to comment objectively on the work of 'Little Endians'. It is not that 'hot' fusion physicists could not be trusted to be impartial, or were incapable of accepting experimental facts, but rather that they would be coming from a research background that would naturally give them a quite different perspective.

There was also the problem of money. Whoever develops a working fusion reactor – hot or cold – will be providing the source of energy that mankind needs for the foreseeable future: perhaps for hundreds of years. The patents involved in the technology, and the head start the patent owners will have in setting up a new power industry, will be worth many billions of pounds in revenue. It is potentially the most lucrative invention ever made. With such big sums at stake, the scientists' university wanted no future ambiguity about who was claiming priority, and hence encouraged them to mount a very public announcement.

In the end, the two scientists agreed to a press conference that would stake Utah University's claim to priority in any future patent applications, followed by publication of a joint paper in their own professional journal, *The Journal of Electroanalytical Chemistry*.[1]

There followed a brief honeymoon of a week or two, during which newspaper libraries received more requests from the newsroom for cuttings on fusion than in the previous twenty years, and optimistic pieces about cheap energy from sea-water (where deuterium is common) were penned to keep features editors happy. All over the world, laboratories raced to confirm the existence of cold fusion, although many scientists were unhappy at the lack of scientific detail and at having to learn about such an important event from television news and the popular press. What these researchers were looking for, with their £90-worth of precious metals stuck in test-tubes, were one or more of the key tell-tale signs that would confirm cold fusion. When two deuterium nuclei fuse they produce either helium and a neutron particle or tritium and a proton. So, if fusion really is taking place, it should be possible to find neutrons being emitted, or helium being formed or tritium being formed. It should also be possible to detect energy being released, probably as heat, that is greatly in excess of any electrical energy being put in. (Of course, if the cell does not do this it is of no use as a power source.)

Despite the experimental difficulties it was not long before confirmations were reported. First were Texas A & M University, who reported excess energy, and Brigham Young University who found both excess heat and measurable neutron flow. Professor Steve Jones of BYU said his team had actually been producing similar results since 1985, but that the power outputs obtained had been microscopically small, too small in fact to be useful as a power source.[2,3]

One month after the announcement the first support from a major research institute came when professor Robert Huggins of California's Stanford University said that he had duplicated the Fleischmann-Pons cell against a control cell containing ordinary water, and had obtained 50 per cent more energy as heat from the fusion cell than was put in as electricity. Huggins gained extra column-inches because he had placed his two reaction vessels in a red plastic picnic cool-box to keep their temperature constant.[4] This kitchen-table flavour to the experiment added even further to the growing discomfort of hot fusion experts, with their billion-dollar research machines.

By the time the American Chemical Association held its annual meeting in Dallas in April 1989, Pons was able to present considerable detail of the experiment to his fellow chemists. The power output from the cell was more than 60 watts per cubic centimetre in the palladium. This is approaching the sort of power output of the fuel rods in a conventional nuclear fission reactor. After the cell had operated from batteries for ten hours producing several watts of power, Pons detected gamma rays with the sort of energy one would expect from gamma radiation produced by fusion. When he turned off the power, the gamma rays stopped too. Pons also told delegates that he had found tritium in the cell, another important sign of fusion taking place.

Pons estimated that the cell gave off 10,000 neutrons per second. This is many times greater than the rate of background level of natural radioactivity, but is still millions or billions of times less than the rate of neutron emission that one would expect from a fusion reaction – a puzzle which Fleischmann and Pons acknowledge as a stumbling block to acceptance of their phenomenon as fusion by any conventional process.[5]

However, despite the reservations, the assembled chemists were ecstatic that two of their number had apparently scooped their traditional rivals from the world of physics, and had, in the words of the American Chemical Society's president, 'come to the rescue of fusion physicists'.

This was perhaps the high-water mark of cold fusion. Scores of organisations over the world were actively working to replicate cold fusion in their laboratories, and although many reported difficulties a decent number reported success. And by the end of April, Fleischmann and Pons were standing before the US House Science, Space and Technology committee asking for a cool $25 million to fund a centre for cold fusion research at Utah University.

Then things began to go wrong. First, some of the researchers who early on announced confirmation of cold fusion now recanted, citing faulty equipment or measurements. Next, an unnamed spokesman for the Harwell research laboratory – the home of institutional nuclear research in Britain – spoke to the *Daily Telegraph* saying that:

> we have not yet had the slightest repetition of the results claimed by professors Martin Fleischmann and Stanley Pons. Of the other laboratories around the world who have tried to replicate the Pons-Fleischmann result, all but one have recanted, admitting that either their equipment or their measurements were faulty.
>
> We believe our experiments are much more careful than those conducted by others. Perhaps for that reason we have been unable to observe any more energy coming out of the experiment than was put in.[6]

And by the time the American Physical Society had *its* annual meeting in Baltimore in May, the opponents of cold fusion were gathering strength. Steven Koonin, a theoretical physicist from the University of California at Santa Barbara, received rapturous applause from the physicists when he declared, 'We are suffering from the incompetence and perhaps delusion of doctors Pons and Fleischmann.'[7]

It was, however, a chemist, Dr Nathan Lewis of the California Institute of Technology, who got the loudest applause. Lewis told the delegates that after exhaustive attempts to duplicate cold fusion, they had found no signs of unusually high heat. Nor did they detect neutrons, tritium, gamma-rays or helium.

By late May, the headlines in both the popular press and the scientific press were beginning to carry words like 'flawed idea' when the biggest blow of all hit supporters of cold fusion. Dr Richard Petrasso of the Plasma Fusion Center of the ultra-prestigious Massachusetts Institute of Technology presented the results of a series of intensive investigations into the Fleischmann-Pons experiment. The fundamental data put forward by the two men, said Petrasso, was probably a 'glitch'. The entire gamma-ray signal in the Fleischmann-Pons experiment, he said, might not have occurred at all.

'We can offer no plausible explanation for the feature other than it is possibly an instrumental artefact with no relation to gamma-ray interaction,' he told the same reporters who had clustered around Fleischmann and Pons only two months earlier.

Dr Ronald Parker, director of MIT's Plasma Fusion Center, said: 'We're asserting that their neutron emission was below what they thought it was, including the possibility that it could have been none at all.'[8]

Thus within two months of its original announcement, cold fusion had been dealt a fatal blow by two of the world's most prestigious nuclear research centres, each receiving millions of pounds a year to fund atomic research. The measure of MIT's success in killing off cold fusion is that still today, the US Department of Energy refuses to fund any research into it while the US Patent Office relies on the MIT report to refuse any patents based on or relating to cold fusion processes even though hundreds have been submitted.

If Dr Parker had left his statement there, it is likely that the world would never have heard of cold fusion again – or not until a new generation of scientists came along. But having been so successful at discrediting MIT's embryonic rival, he decided to go even further and openly accuse Fleischmann and Pons of possible scientific fraud.

According to Dr Eugene Mallove, who worked as chief science writer in MIT's press office, Parker arranged to plant a story with the *Boston Herald* attacking Pons and Fleischmann. The story contained accusations of possible fraud and 'scientific schlock' and caused a considerable fuss in the usually sedate east-coast city. When Parker saw his accusations in cold print and the stir they had caused he backtracked and instructed MIT's press office to issue a press release accusing the journalist who wrote the story, Nick Tate, of misreporting him and denying that he had ever suggested fraud. Unfortunately for Parker, Tate was able to produce his transcripts of the interview which showed that Parker had used the word 'fraud' on a number of occasions.[9]

It then began to become apparent to those inside MIT that the research report that Parker and Petrasso had disclosed to the press in such detail was not quite what it seemed; that some of those in charge at MIT's Plasma Fusion Center had embarked on a deliberate policy of ridiculing cold fusion and that they had – almost incredibly – fudged the results of their own research.

The MIT study announced by Parker and Petrasso contained two sets of graphs. The first showed the result of a duplicate of the Fleischmann-Pons cell and did, indeed, show inexplicable amounts of heat greater than the electrical energy input. The second set were of a control experiment that used exactly the same type of electrodes, but placed in ordinary 'light' water – essentially no different from tap water. The results for the control cell should have been zero – if cold fusion is possible at all, it is conceivable in a jar full of deuterium, but not in a jar of tap water. Any activity here, according to current theory, would simply indicate some kind of chemical, not nuclear, process.

But the MIT results for the control showed exactly the same curve as that of the fusion cell. It was the identical nature of the two sets of results that depicted so graphically to the press and scientific community the baseless nature of the Fleischmann-Pons claim and that justified MIT's statement that it had 'failed to reproduce' those claims. It was these figures that were subsequently used by the Department of Energy to refuse funding for cold fusion and by the US Patent Office to refuse patent applications. And it is these figures that are used around the world to silence supporters of cold fusion.[10]

But MIT insiders, such as Dr Eugene Mallove, were deeply suspicious of the published results. It is usual for experimental data to be manipulated, usually by computer, to compensate for known factors.

No one would have been surprised to learn that MIT had carried out legitimate 'data reduction'. But what they had done was selectively to shift the data obtained from the control experiment, the tap water cell, so that it appeared to be identical to the output from the fusion cell.

When this fudging of the figures became public, MIT came under fire from many directions, including members of its own staff. Eugene Mallove announced his resignation at a public meeting and submitted a letter to MIT accusing them of publishing fudged experimental findings simply to condemn cold fusion.[11] A number of critical papers were published in scientific journals culminating in the paper published by *Fusion Facts* in August 1992 by Dr Mitchell Swartz in which he concluded,

> What constitutes 'data reduction' is sometimes but not always open
> to scientific debate. The application of a low pass filter to an elec-
> trical signal or the cutting in half of a hologram properly constitute
> 'data reduction', but the asymmetric shifting of one curve of a
> paired set is probably not. The removal of the entire steady state
> signal is also not classical 'data reduction'.[12]

In the restrained and diplomatic language of scientific publications this is as close as anyone ever gets to accusing a colleague of outright fiddling of the figures to make them prove the desired conclusion.

Beleaguered and under fire from every quarter (except the other big hot fusion laboratories who simply became invisible and inaudible) MIT backed down. It added a carefully worded technical appendix to the original study discussing the finer points of error analysis in calorimetry. It also amended its earlier finding of 'unable to reproduce Fleischmann-Pons' to 'too insensitive to confirm' – a rather different kettle of fish.[13]

Although MIT changed its story, it was its original conclusion that stuck both in the public memory and as far as public policy was concerned. The *coup de grâce* was delivered to cold fusion when the US House committee formed to examine the claims for cold fusion came down on the side of the skeptics. 'Evidence for the discovery of a new nuclear process termed cold fusion is not persuasive,' said its report. 'No special programmes to establish cold fusion research centers or to support new efforts to find cold fusion are justified.'

Just where does cold fusion stand four years after the original announcement? The position today is that cold fusion has been experimentally reproduced and measured by ninety-two groups in ten countries around the world. Dr Michael McKubre and his team at Stanford Research Institute say they have confirmed Fleischmann-Pons and indeed say they can now produce excess heat experimentally at will. Many other major universities and commercial organisations have also confirmed the reality of cold fusion. US laboratories reporting positive results include the Los Alamos National Laboratory, Oak Ridge National Laboratory (these were the two US research establishments most closely involved in developing the atomic bomb), Naval Research Laboratory, Naval Weapons Center at China Lake, Naval Ocean Systems Center and Texas A & M University. Dr Robert Bush and his colleagues at California Polytechnic Institute have recorded the highest levels of power density for cold fusion, with almost three kilowatts per cubic centimetre. This is thirty times *greater* than the power density of fuel rods in a typical nuclear fission reactor. Overseas organisations include Japan's Hokkaido National University, Osaka National University, the Tokyo Institute of Technology and Nippon Telephone and Telegraph Corporation, which has announced that its three-year research programme has 'undoubtedly' produced direct evidence of cold fusion. Fleischmann and Pons are working for the Japanese-backed Technova Corporation, a commercial cold fusion company based in France. Eugene Mallove left MIT to become editor of *Cold Fusion* magazine.

The Japanese government, through the Ministry of International Trade and Industry (MITI) has announced a five-year plan to invest $25 million in cold fusion research. The Electric Power Research Institute (EPRI) in California has spent some $6 million on cold fusion already and budgeted $12 million for 1992. In addition, a consortium of five major US utility companies have committed some $25 million for EPRI research.[14]

Some of these research funds are being spent not only on developing a large-scale reactor vessel for use in public utilities but also, because of the inherent simplicity and relative safety of cold fusion, the development of a cheap miniature version for use in the office and even in the home. Even as Harwell and MIT proclaim their impossibility, prototype ten Kilowatt cold fusion heating devices are already under test and are likely to find their way to market in the near future.

It is not only the organisations with a vested interest that come out badly from the story of cold fusion. The press, especially the scientific press, has acquitted itself poorly. *Nature* magazine showed how reactionary it can be with coverage that ranged from knee-jerk hostile to near hysterical. Its most intemperate piece was an editorial column in March 1990 headlined 'Farewell (not fond) to cold fusion', which described cold fusion as 'discreditable to the scientific community', 'a shabby example for the young', and 'a serious perversion of the process of science'.[15]

Some sections of the national press were also quick to ridicule Fleischmann and Pons and wrote pieces that have now come back to haunt their consciences. Steve Connor, writing in the *Daily Telegraph*, said that 'The now notorious breakthrough in "cold fusion" only two months ago astonished scientists worldwide, promising a source of limitless energy from a simple reaction in a test tube. Mounting evidence suggests the whole notion is a damp squib.' Connor went on to ask 'how two respected chemists could apparently make such a blunder'?[16] He provides an answer with the suggestion that Fleischmann and Pons were the victims of 'pathological science' – cases where otherwise honest scientists fool themselves with false results.

It is, of course, always fun to read about a good scandal, especially when the detractors who are so free with scorn get their come-uppance so poetically. But the aspect of the cold fusion affair that interests me most is why – exactly why – some scientists felt an overwhelming need to suppress it, even to the extent of behaving in an unscientific way and fudging their results. Money is the most obvious answer, but somehow unsatisfying; they may well have wanted the big research funds to continue to roll in year after year, but that cannot be the whole story. By enthusiastically embracing this possible new field, any of the world's fusion research organisations could have *increased* their research funds, rather than lost anything.

Injured pride is also plausible – men and women are often driven to extremes of behaviour by such feelings, even including murder and suicide. But it is hard to see exactly how and why the feelings of hot fusionists should be so hurt by a simple scientific discovery.

Some interesting clues to this extraordinary behaviour come from examining the reasons that several of the institutions themselves gave publicly for wanting to suppress such research during the development of the affair.

The first sounds perhaps the most reasonable. John Maple, a spokesman for the Joint European Torus project at Culham, Oxfordshire, the world's biggest fusion research centre, told the *Daily Telegraph* that a discredited cold fusion might produce a backlash that would damage the funding prospects of hot fusion.

> People in the street often don't know the difference. They confuse cold fusion, which we think will never produce any useful energy, with the experimental work we are doing at Culham, involving temperatures of hundreds of millions of degrees, which is making spectacular progress.[17]

These sound very understandable fears, but look a little closer at the logic underlying them. The people in the street (that's you and me) 'can't tell the difference'. The difference between what? The difference between hot fusion (which is real) and cold fusion (which John Maple and his colleagues say is not real). But surely, the issue is not whether we, the public, can tell the difference between a nuclear process that is real and one that is not, but whether we, the public, should be asked to entrust millions of pounds of research funds to people who appear resistant to accepting the reality of a process such as cold fusion, for which there is substantial evidence and which may in the long term produce energy far more cheaply than the hot fusion process.

At quite an early stage in the affair, Harwell nuclear research laboratory began to worry about fusion becoming the province of everyman. Members of the public were apparently telephoning Harwell and asking for advice on how to perform cold fusion experiments. 'I have had many odd calls from people', a spokesman told the *Daily Telegraph* in April, 'saying they are going to set it up at home to make it work. One housewife claimed that she already had supplies of heavy water and was asking me for details of how to set up the experiment. I had to tell her it would be extremely unwise.' The paper then costed the experiment at £28 for some platinum, £31 for the palladium, £6 for some lithium chloride and £18 for the heavy water. With a few pounds for batteries, test-tubes and the like, the total could come to as little as £90, leading the paper to suggest that concern was mounting for the 'retired professors, cranks and housewives' who it thought might be joining the race to produce fusion on their kitchen tables.[18]

It is, of course, touching for Harwell to be so concerned about the safety of the man and woman in the street, but I see another worrying part of the explanation in this amusing reaction. Anyone who interests themselves in cold fusion is immediately labelled as belonging to a group that has either lost its marbles or never had any in the first place – 'retired professors, cranks and housewives'. Since we, the people in the street, pay many millions each year to fund Harwell, it seems not unreasonable that members of the public should be able to telephone to enquire on scientific matters without being ridiculed, patronised or told, in effect, to mind their own business.

It was not long before Europe's most senior fusion scientist, Dr Paul Henri Rebut, director of the JET laboratory at Culham (cost, £76 million a year) was offering a word of advice to the man and woman in the street while also, curiously, disclaiming any supernatural powers. 'I am not God, and I don't claim to know everything in the universe. But one thing I am absolutely certain of is that you cannot get a fusion reaction from the methods described by Martin Fleischmann and Stanley Pons.'

Dr Rebut clinched his argument with a single decisive stroke. 'To accept their claims one would have to unlearn all the physics we have learnt in the last century.'[19] Well, we certainly wouldn't want one to have to do that, would we?

Equally illuminating were the remarks of Professor John Huizenga, who was co-chairman of the US Department of Energy's panel on cold fusion and who came down against the reality of the process. In a recent book on the subject, Professor Huizenga observed that 'The world's scientific institutions have probably now squandered between $50 and $100 million on an idea that was absurd to begin with.'[20]

The question is, what were his principal reasons for rejecting cold fusion. Professor Huizenga tells us: 'It is seldom, if ever, true that it is advantageous in science to move into a new discipline without a thorough foundation in the basics of that field.'

When you consider that his committee's sole function was to advise whether or not research funds should be spent to investigate an entirely new area of physics/electrochemistry, and that this statement is one of his principal reasons for deciding *not* to invest such research funds, his remarks take on an almost Kafkaesque quality. It is unwise to invest research funds in any new area, unless we already have a thorough foundation in the basics of that new area? How could anyone ever get any money for research out of Professor Huizenga's committee? By proving that they already know everything there is to know?

Cold fusion is the perfect exemplar of the taboo reaction in science. It runs entirely counter to intuitive expectation produced by the received wisdom of physics; it is a discovery by 'outsiders' with no experience or credentials in fusion research; its very existence is vehemently denied, even though Fleischmann and Pons have demonstrated a jar of water at boiling point to the world's press and television; and it is inexplicable by present theory: it means tearing up part of the road-map of science and starting again – 'unlearning the physics we have learnt.'

But if this discovery is the subject of scorn and derision by orthodox scientists, the subject of the next chapter makes cold fusion seem as solid and well-documented as the law of gravity.

CHAPTER FOUR

The Gates of Unreason

*Will the gates of unreason then be allowed
to open and drown us in a world inhabited by
spirits of the dead and the like?*

JOHN TAYLOR

Science and the Supernatural

One day in the summer of 1975, a number of distinguished scientists and their colleagues crowded expectantly into a 15 foot by 15 foot laboratory at Birkbeck College, University of London. Those packed into the room included Dr John Hasted, professor of experimental physics, Dr David Bohm, Birkbeck's professor of physics, and writer Arthur Koestler. A further seven people, mainly scientific colleagues of Hasted's were also present.

Hasted and his colleagues had been avoiding the press all day and eventually they disconnected the telephone to prevent any interruption of the crucial test they had assembled to witness.[1]

When the test subject arrived, Hasted used a commercial Geiger counter to ensure that he had no radioactive materials on him. The extent of background radiation was checked, too, and those present also confirmed that it was impossible to manufacture a reading from the Geiger counter by roughly handling it or the cable that connected it to a chart recorder.

When the equipment was ready, Hasted handed the Geiger counter to the subject and asked him to produce a reading by paranormal means. After two minutes, the chart recorder registered two large pulses, one of around twenty five counts per second. At the same time, the subject (who could not see the chart recorder) reported feeling a 'shock'. After a further sixteen minutes of concentration there was another pulse registered on the recorder and again after another five minutes, when the subject said he felt a prickly sensation. During the experiment, a second chart recorder, connected to a gaussmeter (an instrument for measuring magnetism), registered two large pulses which happened at the same time as the later pulses on the Geiger counter.

This experiment was repeated again the next day under similar conditions. During a twenty-five-minute session, another four abnormal pulses of around ten counts per second were produced by the subject, on demand. Finally, at the end of this second session, all observers were asked to leave the laboratory except for Hasted and Arthur Koestler, in order that the subject might concentrate fully, and he was asked to make an extraordinary effort to produce an abnormal reading by paranormal means. Within three minutes, the subject produced a reading that was off the scale of the Geiger counter but which, according to Professor Hasted, may have been as high as 200 counts per second.

Many readers will already have guessed that the test subject was Uri Geller. Equally, whether aware of the identity of the subject or not, many people will find this experiment simply impossible to believe, even though it was conducted by a physicist of the highest reputation under carefully controlled conditions at one of Britain's leading centres of research in physics and witnessed by many others of impeccable reputation. What on earth are we – whether scientists or non-scientists – to make of it?

The difficulty for all of us in accepting reports like this, despite their pedigree, is that they raise questions of such fundamental importance as to be frankly frightening in their implications. If it is possible to cause physical phenomena to occur simply by thinking about them or in some way willing them, isn't the entire structure of science invalidated? Would it not inevitably mean that we do not understand the first thing about the true nature of the world in which we live?

And yet, at the same time, few people would disagree that if the phenomena described by Professor Hasted and others are indeed real, then we are dealing with one of the most important scientific discoveries ever made. You might expect the world's scientists to be impatient to put such claims to the test experimentally and to find out empirically whether paranormal phenomena actually occur. Yet the number of institutional scientists accepting this challenge can be counted on the fingers of one hand and the publicly-funded experimental resources at their disposal can be counted on the fingers of the other.

Despite the potential importance of reports of psychokinesis and related paranormal phenomena, institutional science devotes virtually no resources to their research. In the main this is because there is a widespread (and not unhealthy) skepticism which is summed up in the idea 'if there were anything in it, we would have stumbled across major replicable phenomena years ago. If there were money to be made from the paranormal then our supermarket shelves would already be stacked high with telepathic toothpaste and psychic cornflakes.'

This same healthy skepticism makes science feel confident that there are few – if any – big surprises left in a field like inorganic chemistry, where the chemical behaviour of the elements has been minutely studied and is believed to be understood to a very high degree. It seems simply not worth spending large sums on big projects to discover if sodium chloride has any more rewarding property than flavouring our fish and chips. And again, if there were commercial prizes to be gained from such encyclopaedic searches, they would already have been exploited by profit-hungry corporations with huge research and development budgets.

But although this skepticism acts as a valuable restraint on wasted effort and sloppy thinking, it can be – and sometimes has been – badly misplaced in the past and is an unreliable guide to our expectations of future scientific discovery, especially in those areas where our self-confidence is high but our knowledge is actually rather thin.

Giorgio Piccardi, director of the Institute for Physical Chemistry in Florence, became dissatisfied with conventional explanations of common chemical reactions taking place in water. He noticed that the rate at which reactions took place seemed to vary, and sometimes they did not take place at all. His curiosity was further aroused by his discovery that if he enclosed his experiments in copper sheeting, they always worked as theoretically predicted. Wishing to get to the bottom of this mystery, Piccardi and his colleagues in Florence conducted a heroically long series of chemical experiments simply to see how they fluctuated. They chose a very simple chemical reaction – the rate at which bismuth oxychloride formed a cloudy precipitate when poured into distilled water. Over a ten-year period, Piccardi and his assistants conducted this simple reaction more than 200,000 times, recording the time of day and the time for the reaction to take place.

The results, published in 1960, show that variation in the rate at which the reaction took place was related to solar eruptions and changes in the Earth's magnetic field. Over the longer term, the reaction time varied regularly with the eleven-year cycle of sunspot activity. Control experiments, conducted under copper sheeting, remained unaffected by external influences throughout the experiment.[2]

Piccardi's experiments, which strongly suggest that water is susceptible to influence by electromagnetic radiation, have been repeated and confirmed by a team at Brussels University.[3] Further confirmation comes from research carried out by a US team at the Atmospheric Research Center in Colorado, who showed that it is the water, not the other chemicals involved in the reactions, that is sensitive to electromagnetic fields.[4]

One very obvious conclusion to be drawn from such a fundamental discovery is that there is a strong likelihood that all living organisms – whose bodies consist of chemical reactions taking place in water – are in some currently unknown ways capable of being affected by electromagnetism. Until Piccardi's results were published (and, sadly, still today) most physicists and chemists would simply reject any such ideas as superstitious nonsense.

Fundamental discoveries also remain to be made in a field as basic as electromagnetism itself. As recently as 1975, a sixth-form schoolgirl in England wondered in her science class what would happen if she wound an electrical coil with more turns in the centre than at the ends. The result was the chance discovery of the constant-pull solenoid – a basic electro-mechanical device that had somehow eluded scientists and engineers from Faraday onwards but which today is a staple invention incorporated into literally hundreds of domestic and industrial devices (and earning very substantial commercial rewards).

Examples such as these show that there are scientific discoveries of major importance still to be made. They are not hidden deep in the atomic nucleus or in the remote galactic centre: they are here in our everyday lives. They are not so cunningly concealed that they require billion-dollar particle accelerators or teams of radio astronomers to unravel them. They happen every day in plain view and they can be grasped by a sixth-form schoolgirl or a single chemist with an enquiring mind.

In the case of paranormal phenomena such as psychokinesis, or PK – the movement of remote objects without contact – and extra sensory perception, or ESP, the most surprising thing is not the scarcity of hard evidence obtained under strictly controlled conditions but the abundance of such evidence. Although institutional science ignores such research, there have been literally hundreds of experiments conducted in scores of liberal-minded universities and private research laboratories around the world over the past sixty years and the amount of hard data that has been accumulated is staggering – so much data has been obtained, in fact, that paranormal scientists are now conducting studies of studies – more of which later.

Dr John Taylor, professor of mathematics at King's College, London conducted a series of experiments with Uri Geller and other individuals apparently capable of paranormal effects in 1974. His series of experiments, some of which are described below, also showed that inexplicable physical events are taking place, under controlled conditions, and repeatably on demand.

Although he was to undergo a significant change of mind (described later) Taylor's original conclusion in his 1976 book *Superminds* was that:

> Uri Geller appears to have posed a serious challenge for modern scientists. Either a satisfactory explanation must be given for his phenomenon within the framework of accepted scientific knowledge, or science will be found seriously wanting. Since such an explanation appears to some to be impossible, either now or in the future, they argue that the Geller phenomenon is incompatible with scientific truth, and that the value of reason and the scientific point of view is therefore an illusion. Will the gates of unreason then be allowed to open and drown us in a world inhabited by aetheric bodies, extra-terrestrial visitors, spirits of the dead and the like? Will reason then wholly give way to superstition?

One of the very few physicists to speak out publicly and confront this problem head-on, Taylor then concluded that, concept-shattering though it may be, there could be much to gain from a scientific explanation of the powers of Uri Geller. 'Above all,' said Taylor, 'scientists should not shirk this challenge to the way they view the world. Ways must therefore be found of applying the scientific method to discovering the cause of the phenomenon.'

Most of the friends and colleagues with whom I have discussed Uri Geller were united in their reaction to my questions: that Geller is some kind of stage conjuror or magician; that he has been examined only in uncontrolled environments such as television studios; that his effects are not repeatable on demand; that inexperienced scientists are easily fooled by conjuring tricks; that Geller himself has been caught cheating and 'exposed' as a fraud; that there is sparse and unreliable hard evidence for any serious kind of paranormal phenomenon. Strangely, the facts are pretty nearly the exact opposite of these widely held beliefs.

Geller has been recorded on videotape, in controlled laboratory conditions, observed by professional conjurors and physicists, using no materials he has provided or had access to, and has been filmed producing objects out of thin air, affecting scientific instruments remotely, and locating various concealed materials without any error at all.[5]

Geller has been the subject of at least half a dozen different sets of controlled experiments in recognised scientific institutions. The first was at the Stanford Research Institute in California in 1972, by physicists Harold Puthoff and Russell Targ, and again at Stanford in 1973 with the same researchers; with Professor John Taylor and colleagues at King's College in 1974; with Professor John Hasted and colleagues at Birkbeck College in 1975. There have also been experiments conducted in laboratories in France, Japan and elsewhere in the United States. Some of the results that have been obtained with Geller under controlled laboratory conditions are far more convincing than the spoon bending that many people remember from his television appearances.

In November 1972, at Stanford Research Institute, Geller was filmed and videotaped, continuously scrutinised for sleight of hand or conjuring tricks, and the experiments were performed wherever possible on a double-blind basis — neither Geller himself nor the experimenters, Puthoff and Targ, knew the correct answers to the problems they set him to solve paranormally.[6]

In the first experiment, a verified dice was placed inside a sealed box, shaken by one of the experimenters and placed on a table. It was impossible for anyone to see how the dice had fallen, and Geller's task was to guess paranormally. The test was performed ten times. On two occasions, Geller refused to guess, saying he was unable to perceive the result. But on the eight occasions he did guess he was correct in every instance — a 100 per cent success record in a task the odds against which are about one in a million.

In the next test a research assistant placed an object inside one of ten identical aluminium film containers, which were then stood at random on a table. The objects used included water, steel ball-bearings, magnets and a sugar cube. Geller was asked to guess which container the object was in. He performed this test twelve times and was correct twelve times. As before, he refused to guess on two occasions saying that he could not get a clear perception (of the sugar cube and a ball-bearing wrapped in paper). But once again, the odds on the test were a million to one against.

To test his powers of psychokinesis, the physicists used an electronic precision balance, placing a 1 gram weight on its pan, inside an aluminium film can, and then covering the balance with a glass jar. The output voltage from the scale was recorded on a chart recorder. Geller was then asked to alter the reading on the balance paranormally, while under continuous observation to ensure that he did not touch or in any other way interfere with it. He was able to deflect the chart recorder twice, on each occasion for about one-fifth of a second. On the first occasion, the deflection was equivalent to a decrease in weight of the balance pan by about 1.5 grams, and on the second to an increase of about 1 gram. In other words, he first negated the weight entirely, and then doubled it.

It is impossible to tell from the experiment whether Geller actually affected the weight on the balance or whether he affected the electrical circuits of the apparatus or the pen of the chart recorder directly.

Puthoff and Targ's final experiment in the first series was particularly interesting. While the usual precautions against fraud were taken, Geller was filmed, causing the needle of a gaussmeter to deflect a number of times by passing his hands near the instrument, but not touching it. Several times he caused a full-scale deflection of the instrument, which indicates that he had a magnetic field at least half as strong as the Earth's.

Puthoff and Targ say they saw Geller move iron filings on a sheet of paper by passing his hands near them, which also suggests some kind of magnetic effect (although this event was not video recorded).

John Taylor and his colleagues at King's College had an opportunity to study Geller at first hand in early 1974. Like Stanford, they devised tightly controlled laboratory experiments in which they asked Geller to produce paranormal effects such as metal bending, repeatably on demand. But, Taylor later recalled, it was not easy embarking on such research:

> Besides expecting no thanks from either the believers or skeptics, I realized that I should not expect much help from my academic colleagues. Numbers of them had already expressed displeasure at my appearance on the [Television] 'Talk In' programme with Geller and others soon expressed hostility towards my attempts to start an investigation into the phenomena. I also knew that very little financial support would be available from the usual funding bodies. Nor was the necessary apparatus or laboratory space easily available. However, I did find some staunch allies in my own college.[7]

The tests themselves were every bit as remarkable as those conducted earlier. Taylor asked Geller to demonstrate his spoon-bending ability on demand using a teaspoon that he (Taylor) had brought along for the purpose while Taylor continued to hold one end of the spoon. Taylor says that he held the bowl while Geller gently stroked it with one hand. After around twenty seconds, the thinnest part of the stem suddenly became soft and broke in two, the ends rapidly hardening again in less than a second. Of the result, Taylor said:

> Here, under laboratory conditions, we had been able to repeat this quite remarkable experiment. Geller could simply not have surreptitiously applied enough pressure to have brought this about, not to mention the pre-breakage softening of the metal. Nor could the

teaspoon have been tampered with – it had been in my own possession for the past year.

Taylor also reports a number of instances of samples of metal that had been prepared by the King's College metallurgical laboratory and sealed into glass tubes having been bent paranormally in the presence of Geller and two witnesses who observed him continuously to ensure that he could not touch them directly.

Finally, Geller was asked to produce a paranormal reading on a Geiger counter. It is worth quoting the result in full because Taylor's narrative conveys perfectly the atmosphere at such an event.

> The final test was to determine if Geller could produce a deflection on the Geiger counter; this should indicate whether he could produce radioactive radiation. When it was held near him, Geller registered a zero count on the instrument, taking into account the average background rate of about two counts per second produced by cosmic rays coming from outer space. Geller then took the monitor in his own hands and tried to influence the counting rate. We all stood round looking at the dial and listening for the tell-tale tone.
>
> At first nothing happened, but by extreme concentration and an increase in muscular tension associated with a rising pulse rate, the needle deflected to 50 counts per second for a full two seconds, the sound effects heightening the drama of the occasion. By means of a small loudspeaker each count produced a 'pip', and before Geller affected the machine the sound was of a steady 'pip ... pip ... pip' ... In his hands the sound suddenly rose to become a wail, one which usually indicates dangerous radioactive material nearby. When Geller stopped concentrating the wail stopped and the apparent danger with it. This wail was repeated twice more, and then when a deflection of one hundred counts per second was achieved, the wail rose almost to a scream. Between each of these attempts there was an interval of about a minute. A final attempt made the needle deflect to a reading of one thousand counts per second, again lasting for about twelve seconds. This was five hundred times the background rate – the machine was emitting a scream in the process.

Some five years after these words were written, Dr Taylor had undergone a substantial change of view. In his 1980 book *Science and the Supernatural* he says:

> I started my investigations into ESP because I thought there could be something in it. There seemed to be too much evidence brought forward by too many reliable people for it all to disappear. Yet as my investigation proceeded, that is exactly what happened. Every supernatural phenomenon I investigated crumbled to nothing before my gaze . . .
>
> My present attitude seems to be a complete about-face. My public persona over the last five years has been that of the scientist-manque who has dared to investigate the supernatural; an area into which scientists hardly ever dare to tread. I even began by espousing the supernatural, believing abnormal radio-wave emissions by what has become known as 'sensitives' might be the cause of paranormal phenomena. To disarm the obvious criticism let me say quite bluntly that I am not trying to regain any lost scientific 'respectability' by my new position. Only through constant investigations of many sensitives over a number of years have my earlier positive views on the paranormal been required to be so radically revised.[8]

Of course, it is this radical change of view by Dr Taylor that is so interesting. What exactly brought it about? Taylor tells us in the words already quoted that it is no mere career move brought about by pressure from colleagues. Instead, he gives his reasons in some detail. When science faces up to the supernatural, he says, 'it is a case of "electromagnetism or bust". Thus we have to look in detail at the various paranormal phenomena to see if electromagnetism can be used to explain them.' Taylor concludes that there is no evidence for such electromagnetism in conjunction with paranormal phenomena and that therefore they must be illusory.

What of the test subjects themselves – the 'sensitives'? Although certainly the best-known subject, Uri Geller is far from alone in having repeatedly performed such paranormal feats in controlled conditions. In England there are Nicholas Williams, Stephen North, Julie Knowles and a number of juveniles who remain anonymous such as Andrew G. In France there is Jean-Pierre Girard. In Japan there is Masuaki Kiyota and in Russia there are numerous individuals, the best known of which is Nina Kulagina.

Working with English metal benders, John Hasted has devised extensive methods of guarding against conscious or unconscious fraud. He has, for example, implanted microscopic strain gauges in metal specimens linked electrically to a chart recorder to provide a record of the forces imposed on the specimen. He has recorded many instances of stresses being registered simultaneously from three or more gauges, and extensive deformation of the specimen, under circumstances that rule out fraud. In one famous case, a large piece of aluminium was twisted out of shape by Andrew G., a 12-year-old boy, from a distance of thirty feet.

Doctors Charles Crussard and Jean Bouvaist in France have recorded metal bending by Jean-Pierre Girard when the metal was in glass tubes that had been completely sealed under conditions examined by Hasted and others. Working under the auspices of a French commercial metals company, the investigators went to enormous lengths to ensure that the effects they were examining were produced paranormally and not by normal methods. For example, each metal sample was hallmarked so it could not be substituted, and all its dimensions measured accurately before and after bending. The hardness of the metal was tested before and after and the crystalline structure of the metal was examined by taking 'residual strain profiles'. The structure was also examined under the electron microscope and microphotographs taken. In addition, the chemical composition of the metal was examined before and after. These observations revealed a number of structural anomalies such as a local hardening of the kind produced by compression forces of many tons, but apparently originating internally.[9]

Hasted has adopted similar rigorous precautions to rule out fraud. For example, he and the French researchers have been able to get subjects to bend metal rods that it is beyond the strength of any normal person to bend. Crussard has videotaped Jean-Pierre Girard bending a metal rod by gently stroking it, yet producing a bend that requires some three times the strength of a normal person.

Hasted has also reported the phenomenon of a metal bender turning part of a spoon 'as soft as chewing gum' merely by stroking it but under closely controlled conditions that enabled the plastic deformation to be verified by Hasted himself, and where the chemical composition and weight of the spoon was examined before and after. It is possible to soften a metal spoon chemically but only by causing a corrosion that would leave a number of other alterations such as weight loss, and no such changes were detected.

As well as being repeatable and well attested, scientifically observed psychokinetic phenomena are nothing new. Rudi Schneider was an Austrian who, in the 1930s, was subjected to several long series of experiments under controlled conditions, both in this country and in France. For fifteen months in 1930 and 1931, Schneider was investigated by Dr Eugène Osty of Paris in more than ninety sessions. In the latter part of 1932, Schneider was investigated by members of the Society for Psychical Research in twenty-seven sessions in London.

Schneider, and some of those investigating him, used the terminology of the 1930s, which sounds faintly ridiculous to a modern ear (Schneider was described as a 'trance medium' who had a 'control' called 'Olga'). But the important part of the investigations is not Schneider's personal interpretations for the phenomena he produced but the phenomena themselves and the fact that he produced such a large number of paranormal events while under almost ludicrously restrictive controls.

In London, the experiments were carried out by Lord Charles Hope partly at the home of Lady Rayleigh and partly in a private room. An idea of the thoroughness of the preparation can be gained from the fact that Hope kept the only key to the room being used, and that he hired a firm of builders to inspect it for hidden trapdoors and the like.

At the beginning of each session, Schneider would 'go into a trance' and while in this state would manifest a number of phenomena. The following is a description of one session that took place on 11 November 1932. It was given by C.V.C. Herbert, a professional astronomer and Fellow of the Royal Astronomical Society who was present.

> On one occasion . . . an excellent telekinesis took place when I was one of the sitters. I was seated directly opposite the red lamp, which illuminated a small table on which were placed the objects to be moved. The order of the sitters was as follows: Professor Fraser-Harris and Mrs Fraser-Harris (both holding Schneider), Miss

Reutiner, Mr C.C.L. Gregory, Lady Crosfield, Mr Herbert, The Hon. A.C. Strutt, Mrs P. Quilter, Lord Charles Hope; Captain the Hon. Victor Cochrane-Baillie was taking the notes. The table was heard to move slightly, and eventually fell over, coming to rest on its side, with one edge on my right toe. The legs were pointing towards [two curtains in the corner of the room referred to as 'the cabinet']. At this stage the red lamp was turned up, so that the table was clearly visible. While I was watching it intently, it rose off my toes and then descended onto them again. The total movement was of the order of two inches. A little later this movement was repeated. During both these movements, I could see all four legs clearly, and I am positive that nothing touched the table on the 'cabinet' side. It might have been possible to raise the table by an arrangement of fine black threads placed round the legs; but as the movement was a vertical one, this would have involved the existence of some sort of pulley fixed above the table. Such machinery was out of the question, as it would have been impossible to fix it up and remove it again during the sitting. There can have been nothing in the nature of an extending rod, held by Schneider, such as is sometimes used by fraudulent mediums, as, apart from the fact that Schneider was held by Professor and Mrs Fraser-Harris such a structure must have been visible to me. . . . Granting the integrity of the sitters on my left and right hand, it seems to me impossible that the table could have been moved by normal means.[10]

On reading a description of this sort, one's suspicions of some kind of conjuring trick are aroused at once. But consider this description of the precautions that had been taken in advance. Schneider's chair is sited 35 inches from the nearest part of the small table. The table is 15 inches square, 20 inches tall and weighs 6.5 pounds. The initial position of the table was marked with chalk. There are strips of wood nailed to the floor to prevent Schneider's chair from moving and Schneider is physically held by Dr Fraser-Harris, who is sitting opposite him and holding him by the wrists. Fraser-Harris also keeps his feet outside Schneider's feet. Mrs Fraser-Harris sits immediately beside her husband and also has her right hand on Schneider's right hand or between his hands. At the end of the session, Schneider was searched by both Lord Charles Hope and Fraser-Harris.

During the same session a number of other psychokinetic

phenomena were reported by Herbert and by the official note-taker. They included the ringing of small bells attached to the curtains, on demand, and the playing and movement of a small harp.

When Schneider was in Paris being investigated by Dr Eugène Osty, the precautions were even stricter because Dr Osty made extensive use of infra-red detectors coupled to automatic cameras. The precautions were described as follows:

> A projector of infra-red radiation directed a large beam of invisible light, reflected as often as required by a series of plane mirrors, at a photo-electric cell. The latter, by means of a relay, controlled the opening of a big shutter inserted in the ceiling of the *seance*-room. As soon as any object entered into the infra-red beam this shutter opened rapidly and flooded the laboratory with ultra-violet for 1/10th of a second. Moreover the opening of the shutter simultaneously produced the exposure of a camera provided with a quartz lens and taking a photograph at 1/50th or 1/100th of a second. In this way any gesture towards the object, as well as any supernormal displacement of it, itself caused the taking of a photograph, thus registering any attempt at fraud.[11]

During his fourteenth session with Schneider, Osty recorded that on two occasions, two of the four cameras in use were tripped by interruption of the infra-red beam and photographs were taken automatically. But all the negatives showed nothing abnormal. 'In other words,' said Osty, 'the cause of the photographs was non-photographable.'

Osty brought in a great deal more electronic and mechanical instrumentation to study Schneider and made one further discovery that is of great interest – that though Schneider could repeatedly cause the occlusion of the infra-red beam, he could not cause a complete absorption of it: the most he achieved was about 75 per cent. To achieve a partial absorption of an infra-red beam in this way requires a great deal more than conjuring skill; it requires extensive scientific knowledge and suitable apparatus that would have been impossible to conceal.

Despite the existence of a substantial body of physical evidence such as this, serious scientific research into the paranormal remains practically non-existent today. Indeed, one current academic trend is towards the active stifling of such research.

The clearest recent example of this trend is the appointment in 1992 to a senior research fellowship in parapsychology, at Darwin College Cambridge, of an open opponent of such research – theoretical psychologist Dr Nicholas Humphrey. The research fellowship is known as the Perrot–Warrick fund and is worth £25,000 a year for four years. It was set up in 1931 and was endowed by two members of the Society for Psychical Research to be granted to people 'interested in investigating the existence of supernormal powers of cognition or action in human beings and the persistence of the human mind after bodily death'.

Originally, it was set up at Trinity College because Trinity's professor of philosophy, C.D. Broad, was himself actively involved in paranormal research in the 1930s. However, the bequest has become something of a hot potato in more recent years. Previous recipients of the grant include D.J. Ellis who wrote a controversial book *The Mediumship of the Tape Recorder* which looked at how the voices of the dead might be caught on tape, and Carl Sargent, who conducted research into the subject of telepathy, and who became the first person to write a doctoral thesis on parapsychology at Cambridge since the Second World War. Sargent's studies of extra-sensory perception also aroused much controversy.

The committee who administer the Perrot-Warrick fund decided to pass the bequest on to Darwin College because they foresaw trouble with future appointments in so contentious a field. Darwin was not exactly sanguine about the prospect and one fellow, philosophy professor Hugh Mellor, told *The Guardian* that the college looked with horror on becoming associated with 'spooks, ectoplasm and card games.' The secretary of the Perrot-Warrick committee, forensic scientist Professor Donald West, explained in a letter to the *Journal of the Society for Psychical Research* that the opponents at Darwin believed 'The subject matter was just myth and superstition. . . . The only possible interest was to discover why some people could be induced to believe impossible things.'[12] This opposing view is shared by the scientist appointed by Darwin, Dr Humphrey.

It may seem very surprising that a scientist who professes to be a rationalist should adopt a stance that is the very opposite of rational: that Dr Humphrey should prefer to subscribe to the belief that the subject is merely myth and superstition rather than employing scientific methods to discover whether his belief is true or false. But such a position puts Dr Humphrey very much in the mainstream of contemporary scientific thought.

What parapsychology research there is in Britain today is largely privately funded and is confined to a few individuals working in isolation on virtually non-existent budgets and to the few universities that do not feel intellectually sullied by the subject.

Probably the best-known centre of serious university-based research is the Parapsychology Unit of Edinburgh's Psychology Department. The unit was set up in 1985 and is funded by a bequest of £500,000 from writer Arthur Koestler, who had a life-long interest in science and in parascience.

Robert Morris, professor of parapsychology at Edinburgh has spent several years investigating extra-sensory perception: the supposed ability of some individuals to perceive information remotely by means currently unknown. The idea that Professor Morris has focused on is the possibility that many of those who appear to be able to perceive extraordinary phenomena are in fact better observers than the rest of the population.

To test this idea, Morris first selected test subjects for their 'perceptual vigilance' in seeing images by perfectly normal means that most people would fail to perceive consciously at all. To do so, he exposed test subjects to images projected on to a screen for very short periods, below the normal threshold of perception. Some of the pictures were of emotionally charged subjects (spiders, skulls and snakes for instance) among otherwise neutral images. Morris and his team say that their sample of 'perceptually vigilant' individuals were able to identify the emotionally charged images shown subliminally from amongst the neutral ones. But Morris says that these individuals can also identify such images projected on to a wall in another room, out of sight, thus demonstrating an extra-sensory or paranormal ability as well as an unusual acuity when observing by normal means.

In recent decades the trend has been very much towards carefully controlled statistical trials of this sort rather than the more exciting but less easily controlled levitating of tables or bending of cutlery. Thirty-five years ago, Professor Hans Eysenck of London University's Institute of Psychiatry wrote:

> Unless there is a gigantic conspiracy involving some thirty University departments all over the world, and several hundred highly respected scientists in various fields, many of them originally skeptical to the claims of the psychical researchers, the only conclusion that the unbiased observer can come to is that there does exist a small number of people who obtain knowledge existing in other people's minds, or in the outer world, by means as yet unknown to science.

Writing again on the subject of the paranormal in 1982 and referring to his original statement, Professor Eysenck observed that: 'the only revision necessary now would be that the number of people involved is larger than it was then!'[13]

Much else has changed in those thirty-five years. Hundreds of parapsychology experiments have been carried out in scores of laboratories around the world. During the 1970s and 1980s as results from these studies accumulated, it became fashionable for skeptics to dismiss them by pointing out that their results, though positive, were only a little above chance expectation.

The often heated debate between those scientists who think paranormal experiments disclose real phenomena and their critics who believe there is no such thing as the paranormal, might have carried on indefinitely were it not for an innovative breakthrough that radically shifted the experimental perspective and that throws an entirely new light on the data obtained. In the mid-1970s a new approach to the problem was suggested by psychologist Gene Glass of Colorado University. Glass coined the term *meta-analysis* to describe a new way of combining the results of many different parapsychology studies to make the aggregate results statistically significant. So effective has meta-analysis been that Dr Richard Broughton, director of research at the Institute for Parapsychology in North Carolina, has called it the 'controversy killer', and it has been instrumental in converting prominent entrenched critics of the paranormal to acceptance that the experimental results are real.[14]

In his 1991 book, *Parapsychology: The controversial science*, Richard Broughton explains why meta-analysis is so effective. The scientist's task in experiments such as card guessing, he says, is to extract the significant information from the long run of random events – the signal from the noise – and it is the statistician who provides the necessary tools.

> Most basic statistical tests tell the experimenter only if there really is a signal – the experimental effect – among the noise. But these statistical tests do not provide the experimenter with a definitive answer; they only provide an *estimate* of whether there really is something in all the noise. Like all estimates, the more data that go into a statistical estimate the better it will be. The more data that the experimenter can collect, the more likely it is that the statistical tests will detect a signal in it (if there is one to detect).

When signals are strong, says Broughton, very little data is needed. You need only a second or two at the dial of your radio to know that you are tuned in to your local rock-'n'-roll station. When signals are weak, however, the amount of data collected becomes critical: you may need to spend quite a time at the dial just to identify the language of a foreign station.

Much the same is true in parapsychology testing. If you have a subject who can reliably use ESP to tell which way a tossed coin will fall 5 per cent of the time then you will need to conduct many trials to expose this ability. If you did only twenty coin tosses, the person would guess right roughly ten times and of these only one would be due to ESP – the rest would be chance. Even if you did 100 tosses, you would still be in the position of someone trying to identify a radio station having heard only a tiny fraction of a second of broadcast – it would be indistinguishable from the background noise.

If you did many more – say, 1,000 trials – with the hypothetical subject, more than one-third of your tests will provide significant evidence of your subject's ESP ability (so long as he or she can keep up the 5 per cent rate). And if you did as many as 10,000 trials, you would virtually never fail to detect the subject's ESP.

The problem in the past has been that psychokinesis and ESP have rarely been exposed repeatedly because they are very weak effects ranging from less than 1 per cent to only 2 or 3 per cent above what we would expect due to chance. Note that the important point here is not the strength of the effect, it is the number of trials compared with that strength. And what has happened on many occasions over the past fifty years or so is that parapsychology researchers have carried out experiments with a number of trials that is either inadequate or marginal in exposing such weak effects.

Once you get a number of trials that will show up even a very weak effect, then you can get very clear-cut experimental results – in any field, not just in parapsychology. For example, in 1986 a large-scale trial was begun in the United States to see if aspirin can help combat heart disease. What statisticians call the 'effect size' of aspirin is extremely small (only 0.03). Because its effects are so minimal, if the researchers had studied only 3,000 subjects they would have found that aspirin is no better than a placebo. But because they had 22,000 subjects, the effect became very obvious – the experimenters found that there were 45 per cent fewer heart attacks in the experimental group and they felt the effect so pronounced that they could no longer morally withhold aspirin from the control group, and so discontinued the study.

The relevance of all this to the paranormal is that meta-analysis has enabled scientists to take hundreds of small-scale experiments that, on their own, are incapable of exposing weak paranormal abilities, and assemble them into a super-experiment that gives the sort of numbers of test subjects available with the aspirin trials. And when this aggregation of results is done systematically it shows that the 'effect size' of some paranormal abilities is very substantially bigger than that of the effect size for aspirin and heart disease – as much as 0.55 (against 0.03).

The basic methods of meta-analysis are designed to make different experiments that address the same question statistically equivalent, even though they may have involved different experimental techniques, had different numbers of subjects and produced different results. Once the studies have been made equivalent, they can be combined and an overall assessment of the strength of the effect can be made.

There are, naturally, many problems in harmonising so many different experiments carried out over many years and evaluating the results. Not the least is the dilemma dubbed the 'file-drawer problem' by Dr Robert Rosenthal of Harvard University – one of the best-known exponents of meta-analysis. The file drawer referred to is the hypothetical graveyard for parapsychology studies that produced a nil result. It is reasonable to suppose that such studies have been conducted and that the scientists who performed them did not bother to publish them but merely consigned them in disappointment to the 'file drawer'. Critics of the statistical studies of the paranormal have always appealed to the concealment of these hypothetical studies as being the hidden mechanism providing the statistically positive results of paranormal studies that do get published.

But one of the major advantages of meta-analysis is that it provides a sound basis for calculating exactly how many 'file-drawer' studies there would have to be in order to explain the positive results that *have* been published. And the results of this analysis have completely routed even the sternest critics.

In the case of experiments to see if people can influence the fall of dice by psychokinetic means, meta-analysis has shown that there would have to be nearly 18,000 hidden 'file-drawer' studies in order to reduce the results obtained to chance expectation – that is, there would have to be 121 unpublished failed studies for every study that has been published.

Some of the most outstanding results so far have come from meta-analysis of experiments like those carried out by Roger Nelson of the Princeton Engineering Anomalies Research (PEAR) programme at Princeton University. The research was originated by Robert Jahn, former dean of the School of Engineering and Applied Science. At Princeton, researchers have accumulated years of statistical trials on microscopically small psychokinetic effects – known in the jargon of the paranormal business as micro-PK.

Test subjects are asked to try to influence electronic devices whose output should be random, rather like an electronic version of coin tossing. In one test, for instance, the subject sits and watches an electronic counter which accumulates random numbers very quickly and displays a number. The chance result should be 100 and the subjects try to get consistently either higher or lower numbers on the display. In another test they try to push a cascade of polystyrene balls either more to the left or more to the right as they fall into bins.

To rule out any bias in the equipment, each test subject has to try to get psychokinetic effects in three different ways, first in one direction (for example getting high numbers to light on the display) then in the opposite direction (getting low numbers to light) and a 'baseline' test where he or she tries to have no effect at all. All the results are automatically recorded by computer and Princeton has a virtually unbroken record of every test subject and every experiment.

In December 1989 Dean Radin of Princeton's Psychology Department and Roger Nelson of the PEAR lab published a paper on the meta-analysis of micro-PK experiments not, as might be expected, in a parapsychology journal but in the respected physics journal *Foundations of Physics*. Their paper was entitled 'Consciousness-related effects in random physical systems'. In their analysis, Radin and Nelson tracked down 152 reports describing 597 experimental studies and 235 control studies by 68 different investigators involving the influence of consciousness on microelectronic systems.[15]

Radin and Nelson's article was a bombshell for critics who had for years poured scorn on the statistics of parapsychology experiments. They showed that the meta-analysis of all these trials dramatically provided incontestable evidence for micro-PK. For they found that the odds that the overall result arose due to chance was 1 in 10^{35}. This is as close as anyone in the scientific world ever comes to a 'racing certainty'.

Radin and Nelson also calculated the size of the 'file drawer' of unsuccessful and unpublished micro-PK studies that would have to exist to reduce their result to chance expectation. They found the drawer would have to contain *54,000* such studies.

Summarising their achievement, Dr Richard Broughton says:

> Radin and Nelson's meta-analysis demonstrates that the micro-PK results are robust and repeatable. Unless critics want to allege wholesale collusion among more than sixty experimenters or suggest a methodological artifact common to nearly six hundred experiments conducted over three decades, there is no escaping the conclusion that micro-PK effects are indeed possible. Yet Radin and Nelson, in common with most parapsychologists, stop short of

claiming that all is proven. All they ask is that physicists (and psychologists) start taking these data seriously.

Even with evidence for psychokinesis of this kind, physicists and psychologists are understandably still reluctant to start taking the data seriously. For the fundamental question still remains: if Rudi Schneider, Uri Geller and many other people – perhaps even most people – really can bend spoons, read minds and all the rest, how on earth do they do it? Is there even the slightest evidence for a source of biological energy that could possibly accomplish such astounding feats? Perhaps surprisingly, the answer is that there is a mountain of such evidence. And like that examined so far, it is firmly buried in the files labelled 'taboo subjects – not to be researched'.

....

CHAPTER FIVE

Animal Magnetism

My mind is in a state of philosophical doubt
as to animal magnetism.
SAMUEL TAYLOR COLERIDGE
Table Talk

In 1923, a Russian biologist specialising in the study of cell tissue, Professor Alexander Gurwitsch of the First State University of Moscow, carried out an experiment that sparked heated debate in the international scientific community. Gurwitsch was experimenting with onion plants and he noticed that the cells in the tips of onion roots seemed to be dividing in a definite rhythm, rather than randomly in time.

Looking for a physical explanation, Gurwitsch wondered if the rhythmic cell division might be influenced by nearby cells. To find out, he mounted one root tip in a thin glass tube and pointed it at a similar onion root tip, also protected in a tube but with an area exposed to the effects of the first tube. After several hours' exposure, he examined the 'target' root under the microscope and found that cell division was taking place 25 per cent faster in the exposed area. Gurwitsch concluded that some form of radiation was responsible and tested this idea by trying to block the emission. He placed sheets of glass and gelatin between the source and the target roots and the increased cell division was no longer observable. When he used a thin quartz sheet, however, he obtained the increased division as before. Since glass and gelatin are known not to admit ultraviolet radiation while quartz will, Gurwitsch concluded that the rays emitted by the cells of the onion root tip must be in the ultraviolet or shorter wavelength region. As they promoted cell division or mitosis, he called them mitogenic rays. And after further experimentation he later generalised his observation to the finding that all living cells emit mitogenic radiation.[1]

Numerous professional scientists, both before Gurwitsch and after, have claimed to discover some form of specifically biological energy, usually detectable only under marginal or threshold conditions (their claims are examined in more detail later in this chapter). Gurwitsch was rather different from most because a number of other respected scientists claimed to have replicated his findings. In Moscow, a colleague of Gurwitsch was able to increase the budding of yeast cells by more than 25 per cent by exposing them to mitogenic rays from onion roots. Two French researchers in Paris confirmed Gurwitsch's results and, in Germany, two workers at the Siemens and Halske Electric Company laboratory at Berlin also confirmed the effect. Researchers at six separate laboratories, including Munich, Frankfurt, Leningrad and Paris, succeeded in measuring the radiation directly with modified Geiger counters, although a number of others reported that they were unable to measure it.[2]

Soon after, however, the American Association for the Advancement of Science issued a report saying that Gurwitsch's effect was not replicable and strongly suggested that it was all in his imagination.[3] With this announcement, most researchers gave up any further interest in mitogenic rays and – as we shall see in a later chapter – the very idea of such rays came to symbolise self-deluding scientific research in which perfectly honest scientists fall victim to auto-suggestion. Gurwitsch himself continued publishing papers into the 1930s, but his mitogenic rays had been pronounced imaginary by the AAAS and hence had become a taboo subject.

Matters rested there until 1972 when an extraordinary experiment was reported by S.P. Shchurin and two colleagues from the Institute of Clinical and Experimental Medicine in Novosibirsk, Russia. The experimenters placed identical tissue cultures inside hermetically sealed vessels that were separated by a wall of glass. They next introduced a lethal virus into one of the culture vessels with the result that its tissue colony was destroyed by the virus while the twin colony remained unharmed. When, however, they replaced the glass dividing wall by a quartz divider, they found that not only did the infected cell colony die off, but so did the second colony, even though it was impossible for the virus to penetrate the barrier directly.

The explanation, they concluded, was that the virally infected cells were somehow able to communicate with the uninfected cells, that the communication was in the ultraviolet region of the spectrum, and that it carried information that somehow resulted in the death of the uninfected culture. Shchurin's team repeated the experiments, this time using an electronic detector sensitive to ultraviolet radiation connected to an automatic chart recorder which traced a graph of the ultraviolet emissions. They found that while cell-division processes in the tissue cultures remained normal, the level of ultraviolet radiation registered on the chart recorder remained stable. When the first colony began to die through viral infection, the level of ultra-violet radiation increased. Importantly, the effects were registered and recorded by means of instrumentation, not by human hand and eye.

Shchurin accounted for the unexpected results by saying that since the second tissue colony was dying in exactly the same way as the infected tissue, then it was just as dangerous for the healthy cells to be exposed to the radiation emitted by dying cells as it was for them to be exposed to the virus itself. It was as though the healthy colony, on learning of the death of the infected colony, began to mobilise its defences as if it were being directly attacked, and that this mobilisation was proving as fatal as if it had really been attacked. According to Shchurin:

> We are convinced that the radiation is capable of giving the first warning about the beginning of malignant regeneration and of revealing the presence of particular viruses.[4]

If Shchurin is correct, then his discovery is potentially the most important medical discovery since Pasteur. Yet the malediction of the American Association for the Advancement of Science still hangs over Gurwitsch and any other work that smacks of the self-delusory mitogenic rays, and so Shchurin's research has been ignored in the West.

Future historians of science looking back over the past two centuries may detect an odd recurrent phenomenon. Every fifty years or so from the middle of the eighteenth century, an individual – often a distinguished but eccentric scientist – has claimed to make the same discovery of fundamental importance to humanity; he provides evidence and case studies of his discovery, usually producing a book or paper describing the phenomenon in great detail. He is then ridiculed and ostracised by his profession, sometimes ending his days in rather sorry circumstances. They have dabbled in a taboo subject – one which neither their scientific colleagues nor the rest of the community is willing to hear about.

The discovery I mean is the alleged existence of a specifically biological form of energy; and its propagation in association with living organisms as a biological energy field (bioenergy field for short). Much has been written about such a field: that it is electromagnetic in origin; that it is visible to certain people or under certain circumstances; that it is closely connected with important metabolic processes, especially the immune system, and an important indicator of state of health. In some eastern countries, entire systems of knowledge have been constructed around this central idea. From India across the Himalayas to China, both spiritual and medical philosophy have grown up around belief in bioenergy; systems including Tantric Buddhism, Kundalini yoga and Taoist medicine.

By contrast, the existence of such a field is rejected entirely by orthodox western science and those who have dabbled in the subject have usually become associated in the public mind with crank beliefs of one sort or another.

As well as Gurwitsch and Shchurin, the individuals whose stories I want to take a fresh look at here are Franz Mesmer, Karl Reichenbach, Walter Kilner, Wilhelm Reich and Semyon Kirlian. The first individual on this list of scientific outsiders, Franz Anton Mesmer, enjoys the doubtful distinction of having been maligned by the scientific community not once but twice, his name being inextricably linked both with hypnotism – a practice shunned then and now by orthodox medicine – and with charlatanism. Strangely, although his name is synonymous with hypnotism, there is little historical or scientific evidence for such an association, and evidence to support the charge of extracting cash from gullible and wealthy patrons is, at best, circumstantial.

Mesmer was born in Germany in 1734. He studied at the University of Vienna, becoming a doctor of medicine in 1766 with a thesis entitled *De planetarium influxu*, which he himself later translated into English as *On the Influence of the Planets on the Human Body*. In 1773 he first began to treat patients with magnetic plates that had been invented by a professor of astronomy at Vienna, a Hungarian Jesuit priest called Father Maximilian Hell. A number of patients were reported as being cured by this therapy. From 1776 onwards, he ceased to use magnets directly in his treatment but said that he had discovered a new fluid analogous with, but independent from, mineral magnetism which could affect living organisms – 'animal magnetism'. From Mesmer's writings at the time and later, and from contemporary accounts of his cures – both by himself and by his enemies – it is clear that Mesmer's clinical practices had nothing to do with what we call hypnotism (except in a trivial, superficial way) but were based on a belief in a biological form of energy.

The progression of Mesmer's thinking is also clear from his work and his writing. He began with the idea that the human body is influenced in important ways by the Sun and Moon, an idea we now know to be perfectly correct.[5]

He went from there to the belief that the human body could be directly influenced by magnetic fields, an idea for which there is strong evidence.[6] He finally came to the idea that living things could be influenced by a force field that was analogous to the magnetic field that affected ferrous metals but was actually different – an organic or animal form of magnetism. Although animal magnetism was different from mineral magnetism, he felt justified in drawing a parallel between the two because both were linked with electricity in some way unknown to eighteenth-century science, and he continued to use electrical processes in his therapy.

Between 1773 and 1778 he travelled around Europe demonstrating his new methods in various places from Switzerland to Bavaria. He wrote an account of his discoveries and sent it to the Royal Society in London, the Académie des Sciences in Paris and the Academy at Berlin. Only Berlin bothered to reply and that was to tell him his discovery was an illusion.

In 1778 he arrived at Paris and set up in practice. Over the next few years his fame and the number of patients he treated grew rapidly. He met and befriended Charles Deslon, Doctor Regent of the Faculty of Paris and physician to the Count d'Artois. Deslon was impressed by Mesmer's methods and became first his student and later a partner in practice. Deslon also wrote an account of their work and from him we learn that in 1780 there were seventy people being treated, a waiting list of 600, and a total of several thousand applications from Parisians keen to be treated. A few years later – in 1784 – Mesmer and Deslon are said to have treated no fewer than 8,000 patients in a year, something of a record even by today's conveyor-belt treatment standards.[7]

We do have accounts of a number of specific cures, principally those written up by Deslon, although the sort of details given and the terminology used are such that it is probably impossible now to say exactly what ailments these patients suffered from and what relief they gained by being treated through animal magnetism. However, as Frank Podmore (a writer hostile to Mesmer) observes in his book on animal magnetism, 'It is incredible that the fame of his treatment could have persisted and increased unless many of these persons had derived substantial benefit from it.'[8]

In 1784, a commission from the Royal Academy of Sciences and the Faculty of Medicine was appointed by King Louis XVI to investigate Mesmer and his cures in response to complaints from French physicians that Mesmer was a charlatan. It included such distinguished names as Benjamin Franklin, then in Paris seeking French assistance for the War of Independence, and Antoine Lavoisier, discoverer of oxygen and the nature of combustion. The report was unfavourable and attributed the cures effected to the imagination of the patients (an opinion that was to be echoed many times over the next 200 years). Although the commission found against him, Mesmer continued practising in Paris for some time. Eventually it was the French Revolution that forced him to flee to London, where he remained for some years.

Mesmer's real crime was that he was too successful. His clientele was drawn principally from the upper middle classes: public and court officials, landed proprietors, army officers, priests and even physicians. And his great success was diverting substantial revenue from Paris's medical establishment.

The exact nature of his therapy and whether it was effective has remained a matter of controversy, and it is probably impossible to form a clear judgement today. However, I believe there is one important conclusion that can be drawn from the evidence we have: that it is an almost wilful act of misunderstanding that anyone should interpret Mesmer's actions as having anything to do with hypnotism, when it was so plainly, both expressly and implicitly, concerned with a tangible organic form of force or influence – 'magnetic' being a perfectly natural and logical term for an eighteenth-century scientist to choose to describe a field phenomenon of unknown origin. Mesmer himself expressly differentiated conventional magnetism from the force he sought to describe because he gave it a separate name, *animal* magnetism, and he acknowledged its generic similarity to mineral magnetism.

Although Mesmer did not practise hypnotism, his name became associated with it quite soon after his death and remained synonymously linked for more than a century. The real subject of his experiments, bioenergy or animal magnetism, remained a taboo subject until half a century later when, in 1844, it attracted the attention of one of Germany's most prominent young physical chemists, Karl Reichenbach.

Reichenbach had made his reputation in Germany's flourishing embryonic chemical industry. He was the first to isolate paraffin oil – the most significant form of lighting fuel in the nineteenth century – and he was among the first to make a reputation through experimenting with materials derived from coal tar (he also discovered creosote in this way). He was a competent metallurgist and was among the earliest researchers to take a serious interest in meteorites. However, as far as most scientific histories and reference books are concerned, Reichenbach is a non-person.

In 1844, Reichenbach was introduced by a Viennese surgeon to one of his patients, a hypersensitive neurasthenic young woman of 25, Maria Novotny, who appeared to be able to perceive the field surrounding a very powerful magnet. To rule out suggestion, Reichenbach had his assistant go into the next room and uncover a large magnet directly behind the girl's bed, at which she became uncomfortable and declared there was a magnet around somewhere. She also proved to be able to tell when a magnet had the armature removed from its poles while blindfolded. On a separate occasion when Miss Novotny was unconscious in a cataleptic state, Reichenbach says that a magnet brought near to her hand stuck to the flesh as though to a piece of iron. As Miss Novotny recovered her health, she appeared to lose her paranormal abilities. Reichenbach subsequently looked for and found four other neurasthenic young women who also seemed to be able to perceive magnetic fields.

Like Mesmer before him, Reichenbach at first came to the conclusion that the force he was dealing with was magnetism. He tested experimentally a number of stories he had heard and rejected them. For example, he was told that certain 'sick sensitive' girls (the term he used to describe them) were able to magnetise needles by holding and stroking them. He investigated this claim but says that he found it to be false.

Looking for an alternative explanation he tried the effect of metals other than iron (copper and zinc) and non-metallic crystals (sulphur and alum) on the neurasthenic patients, and found they said they were able to perceive colours associated with these crystals and experienced subjective sensations associated with their touch, as with magnets. Reichenbach concluded that the force he was experimenting with was not magnetism but a field permeating crystalline structures. It was particularly strongly associated with living things. He called it 'odyle' or the odic force.[9]

In 1845, he produced a book detailing his researches, which caused something of a sensation in the German-speaking world. In 1850, the book appeared in English translation with the title *Researches on Magnetism etc. in Relation to the Vital Force*. The book was translated into English by Dr William Gregory, professor of chemistry at Edinburgh University. Reichenbach's principal opponent in England was Dr James Braid, the man who had put the word hypnotism into the English language and who had a simple explanation for mystical Germans and their 'sick sensitives'. It was, said Braid, hypnotic suggestion: a solution with a familiar ring.

When his book received a frosty reception in scientific circles, Reichenbach decided to abandon paranormal research and return to chemistry and metallurgy, which he did with considerable distinction, although too late to save his reputation.

Half a century later, an English doctor stumbled across the same trail, and with very similar results. Walter John Kilner joined the staff of London's principal teaching hospital, St Thomas's, in 1869 as a physician and surgeon. Kilner's name will be familiar in practically every English kitchen since he was the inventor of the sterile preserving jar that bears his name. In the course of his work he started to take an interest in using electrical treatments. When Roentgen discovered X-rays, St Thomas's was quick to establish an X-ray department and in 1897 Walter Kilner was appointed its director.

Kilner started experimenting to try to see in parts of the spectrum that are normally invisible to the human eye. In particular he experimented with making glass screens coated in a chemical called dicyanin, or cyanine blue, which is used in the photographic industry to sensitise photographic emulsions. The chemical makes photographic film sensitive to all parts of the visible spectrum instead of merely a narrow band. Cyanine also occurs naturally in many plants and is responsible for the purplish blue colour of many flowers that also radiate in the ultraviolet region, invisible to the human eye but not to that of many pollinating insects.

Using glass cells containing dicyanin, Kilner claimed to be able to perceive what clairvoyants and occultists refer to as the human aura, but which Kilner – wishing to dissociate himself from the occult world – called 'the human atmosphere'. In 1912, Kilner published a book with this title.[10] In it he detailed his account of experiments in which he viewed the human atmosphere as a faint grey outline surrounding the body, which enabled him to diagnose the state of health or otherwise of some patients.

Publishing the book proved to be an act of professional suicide. The *British Medical Journal* for 6 January 1912 reviewed the book, saying, 'Dr Kilner has failed to convince us that his aura is more real than Macbeth's visionary dagger'. Kilner immediately resigned his position and retired to Bury St Edmunds where he helped his brother in private practice. He died in 1920.

If any one individual can be said to have been the natural heir to Mesmer, Reichenbach and all those who had experimented with bioenergy, it must surely be Wilhelm Reich. Reich has the unique distinction of having been persecuted and his books burned not once, but twice – first by the Nazis in the 1930s, and second by the Americans, in the 1950s.

There are two distinct phases to Reich's career, and they have a by now familiar ring. Initially he trained as a doctor in Vienna, later specialising in psychotherapy and becoming an influential member of Freud's circle and the psychoanalytic movement while still a young man. Freud was keen that psychoanalysis should be a respectable discipline that became acceptable to the medical establishment. In the 1920s, Reich upset the applecart in several ways: by joining the communist party; by espousing free love; and by setting up street clinics offering sex therapy to working-class unmarried men and women. He also fell out with Freud over matters of doctrine.

Reich saw all the processes of the body and mind as processes of bioenergy, ebbing and flowing. The supreme example of this energy flux, he thought, was the orgasm, a function which was invariably impaired in the neurotic people he treated. His criterion for a return to full mental and emotional health was regaining the ability to surrender fully to orgasm, and the inability to do so, he conceived of as a pathological symptom. In 1938–9 Reich began to believe that the bioenergetic processes he was seeing played out in his consulting rooms were not merely emotional or psychological processes but physical processes involving a specifically biological form of energy, a form which he called orgone. He began to experiment with orgone, devising boxes he called accumulators.

Reich quit Germany, first for Norway and then America, just as the Nazis began burning his books. His crimes covered a long list: he was a Jew; he was a communist; he talked about orgasms; strange things went on in his consulting rooms – and he believed in biological energy. Outside Germany, Reich attracted many students and patients who later became influential therapists in their own right.

In the United States his theorising entered a final stage, he parted company entirely with orthodoxy and developed the idea that orgone energy is not merely contained inside living things but is dispersed everywhere, on earth, in the sky and probably in space too. It was, he said, contained in microscopically small vesicles which he called bions. Reich attempted obsessively for some years to interest prominent physicists (such as Einstein) and biologists in the existence of bions, but without success.

He also spread the gospel of his black box orgone accumulators, which he sold by mail order. In 1956, the Federal Food and Drug Administration ruled that his orgone accumulators were a fraud and that he must stop selling them. The FDA ordered that all remaining orgone accumulators and their accompanying instructions must be burned in the public incinerator. The FDA was so determined to protect the American people from corruption that the injunction it obtained from the Federal Court also contained the clause:

> That all copies of the following items of written, printed or graphic matter, and their covers, if any, which items have constituted labelling of the article or device, and which contain statements and representations pertaining to the existence of orgone energy, its collection by, and accumulation in, orgone energy accumulators, and the use of such alleged orgone energy by employing said accumulators in the cure, mitigation, treatment and prevention of disease, symptoms and conditions . . . shall be withheld by the defendants and not again employed as labelling; in the event however, such statements and representations, and any other allied materials are deleted, such publications may be used by the defendants.[11]

The FDA injunction actually listed Reich's books that had to have all references to bioenergy deleted, including such dangerous titles as *The Sexual Revolution*, *The Mass Psychology of Fascism* and *People in Trouble*. In executing the injunction, the FDA's officials chose to interpret it as meaning that all Reich's books mentioning orgone must be burned, and so a medieval spectacle was enacted on a modern American High Street, as lorry loads of books such as *Character Analysis* – one of the most influential contributions to psychotherapy literature of the twentieth century – were consigned to the public incinerator. Reich refused to comply with the injunction, was imprisoned and died the same year in jail, possibly of a heart attack. The FDA continued to burn his books until 1960.

There are certain recognisable characteristics of the unconscious rejection of ideas concerning bioenergy, all of which are illustrated in the cases described here. First there is the deliberate misidentification of the real subject of research. It is usually identified as hypnotism, faith healing, spiritualism, mediumistic trance, clairvoyance, possession and the like. Interestingly, these ideas appear to shade off subconsciously towards an almost dogmatic medieval horror and accusations that border on witchcraft – distant echoes of both the Inquisition and Salem.

There is the people's champion: the right-minded scientist, or public official, who cannot bear to see the community misled by such charlatanism and who feels it is his public duty to speak out, and conduct counter experiments, or provide evidence from conventional scientific sources that show the charlatan up for what he is. His pen is not merely mightier than the sword – it *is* a sword, with which he strikes down the upstart outsider and his weird antisocial ideas.

And, of course, there is the explanation of the so-called paranormal events themselves. Here we are on familiar ground. The explanation of the data – regardless of whether it is on videotape, or occurred in front of multiple witnesses of unimpeachable integrity – has remained the same from Louis XIV's Commission in 1784 until the Federal Food and Drug Administration imprisoned Reich and burnt his books in 1956. It is a simple case of self-delusion on the part of the scientist and imagination on the part of his victims. The spoons that bend, the objects that levitate, the instruments that register forces that no one can see, are simply fairground conjuring tricks that stage magicians can duplicate at will. All are agreed there is no evidence for a bio-energy field.

In 1939, the same year that Reich claimed to have discovered bioenergy in the form of orgone, a Russian scientist, Semyon Kirlian discovered a novel form of photography that, for the first time, appeared to show unmistakable evidence of just such a field – a field that Russian scientists called bioplasmic energy.[12] Kirlian's invention makes use of the Tesla coil – a high-voltage, high-frequency electrical device – to photograph living tissues. Experimenters have used frequencies between 20,000 and 3,000,000 cycles per second, and voltages between 20,000 and 50,000 volts, to generate electric fields that cause electrons to be pulled out of the surface of living tissue such as human skin. The process is not destructive or painful but it does enable the pattern of electrons, in turn, to generate light that can expose a photographic plate. There is nothing mystical or magical about the technique itself. It is a phenomenon called corona discharge and is sometimes seen as a faint glow around high-voltage power lines.

Several US laboratories have taken up Kirlian photography including Dr Thelma Moss of UCLA and Henry Monteith of the University of New Mexico, and have successfully duplicated its effects. Dr Moss wrote that, 'We were able to corroborate what the Soviet literature had reported, that different states of emotional or physiological arousal in the human being would reveal very different patternings in these photographs.'[13]

Just what these photographs show, though, is a matter of some controversy. Proponents of Kirlian's process say that they show quite simply the human aura or bioenergy field. They point to examples of photographs – of which many thousands have been taken – showing such things as the fingertips of a healthy person (which show flares of energy emanating from the skin) and those of a tired or sick person (where the flares are much reduced). Some of Kirlian's photographed results appear strikingly to confirm the claims made by the researchers mentioned earlier. The most famous of all is a photograph of a leaf. When the leaf has been freshly picked from a tree, it radiates flares all around its edge and its surface is covered in brilliant sparkles. When photographed again some hours later, the flares have died down and the surface sparkles have almost died away. Hours later again, the leaf appears completely lifeless.

Even more remarkable, is the related but separate phenomenon of the 'phantom leaf'. If a freshly picked leaf is cut in half, a Kirlian photograph clearly shows the complete original space occupied by the leaf when alive and whole, strongly suggesting that the organism of the leaf is informed by what even orthodox biologists have started calling a 'morphogenetic field'.

Those sceptical of the technique say that it is impossible to know with certainty what the Kirlian photographs show: that the process being recorded is a secondary one and not a direct measurement of any biological field. Kirlian photography was one of the 'paranormal' phenomena examined by Dr John Taylor and his colleagues at King's College and found by them to be 'irrelevant'. Taylor wrote:

These results were used to support the idea of the existence of a 'bioplasma body' or 'etheric body' around a living organism. That inanimate objects could be observed by this technique led to the idea of stones etc., having their own 'etheric bodies'.

Careful work has now been done to test the various claims put forward by believers in the efficacy of Kirlian photography. Two different groups in America, one at Drexel University the other at Stanford University, as well as my own group at King's College, including Eduardo Balanovski and Ray Ibrahim, have attempted to perform Kirlian photography under as well-controlled conditions as possible.

When all [the] factors are carefully controlled there is no change of the Kirlian photograph with the psychological state of the subject. It seems that the most important variable is the moisture content of the fingertip. We can only conclude that Kirlian photography is irrelevant to paranormal investigations.[14]

Remarkably, however, although Dr Taylor's experiments were carried out with the care and attention designed to get to the bottom of Kirlian photography, for some reason not altogether clear, the King's College team chose to ignore the only Kirlian phenomenon worth seriously investigating: the persistence of a 'phantom leaf' shape when part of the leaf body is removed.

One researcher has attempted to deal seriously with this issue. Ion Dumitrescu has challenged the 'phantom leaf' effect, saying:

If a small portion of a fresh leaf is put to one side, the stomata will close, so as to reduce water loss. By placing the leaf onto the Kirlian camera and pressing it onto one of the electrodes a number of microscopic water droplets will be squeezed out a distance of a few millimetres. These microscopic water droplets become, when applying a voltage, centres for causing discharges on the photographic plate, therefore the cut portion also appears in the image.'[15]

Dumitrescu's theory is ingenious but fails to explain the observed facts. The microscopic water droplets said to be responsible would have to be squeezed out five to ten centimetres, not a few millimetres, and they would have to adopt a very specific leaf shape by random chance (they would have to adopt an oak leaf shape, elm leaf shape, or other shape corresponding to the parent leaf by chance alone).

According to British Kirlian researchers Harry Oldfield and Roger Coghill, in their book *The Dark Side of the Brain*:

> Such an explanation is not enough to convince us; an examination of Thelma Moss's photograph of the phantom leaf effect, no less than our own, shows a distinctive outline of the original leaf, and not a secondary discharge uniquely related to any water drops which may have been squeezed out on to the plates.[16]

Semyon Kirlian may turn out to be the first person in history to dabble in the taboo subject of biological energy and not be ridiculed and ostracised. His critics have so far been unable to deploy the conventional argument that his results exist only in the imagination, and have resorted instead to the next most common line of defence, dismissal. So far the strategy has worked perfectly. Kirlian and his discovery continue to be ignored.

But if the experiments of all these researchers do indeed point to the existence of an electromagnetic field or radiation associated with living cells, what exactly are the implications of such a discovery? What, for example, is the relationship of such a field to the body and its metabolism? And what, if anything, are the implications of such a field for medical research?

CHAPTER SIX

A Case of Ill Treatment

Nature never deceives us; it is always
we who deceive ourselves.

JEAN JACQUES ROUSSEAU

When America's National Aeronautics and Space Administration sent teams of astronauts to live in Earth orbit in the Skylab space station in the 1970s, it hoped to gain answers to many urgent questions in space medicine. Not even the most imaginative NASA scientist, however, can have expected to find the first concrete evidence of a causal link between psychological states and the functioning of the human metabolism – the connection between mind and body.

The Skylab programme involved prolonged orbital flights by three successive teams of astronauts between May 1973 and February 1974. Astronauts were in space for a total of 84 days – longer than ever before – and the flights were designed to enable ground-based specialists to monitor the health of people in space in much more detail than on previous missions. One of NASA's principal discoveries was that on the day they were due to return to Earth, and were hence exposed to very high levels of emotional stress, the astronauts' immune systems were affected. Important processes in the immune system such as white cell transformation were abnormally depressed.

This surprise discovery was followed up with clinical trials on Earth, and in 1977 the *Lancet* published the results of an Australian study of the effects of severe emotional stress on the body's immune system. Twenty-six bereaved spouses took part in the study and were examined two weeks after bereavement and again six weeks later. The examinations measured important indicators of immune system function, including hormone concentrations and the number and function of white cells, and these were compared with a control group of healthy people. The results showed that there was a 'striking difference' between the bereaved group and control group in the function known as lymphocyte (white cell) transformation.[1]

Orthodox medical scientists are understandably cautious about this evidence because it has stimulated some unrealistic claims about the benefits of 'mind over matter' type treatments. For some time, it was said by some enthusiasts that as stress caused depression of the immune system then there was a direct link between emotional state and illness. It was thought, for instance, that depression might be a factor in promoting diseases such as cancer. However, further research has shown that this is an oversimplification. Psychiatrist Denis Darko of the Veterans Administration Medical Center in San Diego studied a group of forty-three men suffering severe depression. He found that the *least* depressed men showed the most affected white cell function, while the *most* depressed men compared favourably with the healthy control sample.[2]

Bruce Bower, who reviewed the evidence for and against a link between psychological factors and immune function in *Science News* in 1991, concluded that:

> Some researchers say studies of stress, depression and immunity contain flaws that render them inconsistent and inconclusive. Others see the data in a better light but acknowledge that depression may have received premature billing as a powerful immunity-buster. And everyone admits that so far no solid evidence connects psychological states to any specific immune disease.[3]

This last point is perhaps the most important. Despite evidence of a lowering of immune system functions, no one has specifically linked such changes to catching a *particular* disease. However, the fact remains that the evidence shows that some people are more vulnerable to illness and death immediately after being bereaved – a phenomenon that medical science must explain, rather than ignore.[4]

One reason that orthodox medical scientists have so much trouble with studies such as this is that they strongly suggest a causal connection between state of mind and physical health. Yet this resistance is surprising when you consider just how strong the experimental evidence for such a link is.

Professor Hans Eysenck of London University's Institute of Psychiatry has conducted several major studies in this area. That published in the *British Journal of Medical Psychology* in 1988 is perhaps the best known. It set out to determine whether there is a link between personality, stress and cancer. Over a ten-year period, the study first measured the personalities of the test subjects by certain criteria and followed up later those that died to see if personality type, stress or smoking gave any indication of the likelihood of death from cancer.

> It was found that personality variables were much more predictive of death from cancer or [heart disease] than was smoking, and that different personality types were susceptible to either of these two diseases. Personality type was defined in terms of differential ways of dealing with interpersonal stress, and it was found that stress was a very potent cause of death, in the sense that stressed [subjects] had a 40 per cent higher death rate than non-stressed [subjects].[5]

In 1989, a study published in the *Lancet* reported on the effects of giving psychotherapy counselling to patients with secondary breast cancer. Researchers looked at the effects on survival time of fifty patients who were given weekly sessions of supportive group therapy and lessons in self-hypnosis against pain, compared with a control group of thirty-six patients given normal treatment. The finding was that the experimental group survived an average of three years compared with the control group whose average survival time was only eighteen months.

In this highly significant study the researchers were at pains to make clear that they did not indulge in any 'mystical' practices.

> We intended, in particular, to examine the often overstated claims made by those who teach cancer patients that the right mental attitude will help to conquer the disease. In these interventions patients often devote much time and energy to creating images of their immune cells defeating the cancer cells. At no time did we take such an approach. The emphasis in our programme was on living as fully as possible, improving communications with family members and doctors, facing and mastering fears about death and dying, and controlling pain and other symptoms. To the extent that this intervention influenced the course of the disease, it did not do so

because of any intention on the part of the therapists or the patients that their participation would affect survival time.[6]

The authors cite a dozen other papers from the scientific medical press dealing with the beneficial effects of psychotherapy counselling of cancer patients.

But it is not necessary to go into Earth orbit or to make microscopic measurements of exotic cellular functions to find hard evidence of the link between lifestyle, emotional state and health. In response to the government's Green Paper on the future of the health service entitled *The Health of the Nation*[7], the British Holistic Medical Association published its own consultative document in October 1991.[8] The contents of this document are thought provoking in many of the individual areas of holistic medicine, and none more so than that by Dr Peter Nixon, a consulting cardiologist at Charing Cross Hospital. Nixon points out that:

> The major risk factors for [heart] disease include maternal mal-nutrition and poverty (Barker, 1991), growing up in poor areas (Lord Taylor, 1975), failure at school (Jenkins, 1978), over-whelming burdens with lack of support and appreciation (Cassell, 1977), loss of prediction and control of life events (Weiss, 1972) and low status in a hierarchy of civil servants (Marmot *et al.*, 1978).

Nixon goes on to say that: 'Conventional risk factors of cigarette smoking, raised blood cholesterol and blood pressure, overweight, obesity and lack of exercise, the focus of health education today, may not be prime movers but the outcome of an upbringing which pro-vides a poor capacity for coping with life's hassles and adapting to change. The natural consequence would be a career with high levels of effort and distress, raised levels of stress hormones and increased blood coagulability. The current health education policies do not accom-modate the fact that the [blood protein] level is probably the strongest biochemical index of [heart] disease, and the one with an acknow-ledged psychosocial contribution (Meade, 1989).'[9]

Over the past decade British and American medical journals have carried reports that amount to substantial research evidence that holistic methods of many kinds can be very effective treatments for some diseases. The kind of treatments covered by current medical research include hypnotherapy, acupuncture, acupressure and various forms of psychotherapy counselling, meditation and stress management. Many such treatments are rejected because they appear to call upon either or both of two taboo subjects: a mind-body link whereby physiological healing processes are influenced by psychological factors; and the existence of biological energy paths within the body. Consider the following examples.

In August 1988, the *Journal of the Royal Society of Medicine* carried a paper by a team from the department of anaesthetics at The Queen's University of Belfast on using acupressure to reduce morning sickness in pregnant women. Morning sickness affects up to 88 per cent of pregnant women, most commonly between the sixth and fourteenth weeks (although some women are sick for almost the entire pregnancy.) The drug Thalidomide, whose side-effects were so tragic, was an attempt to control this problem and there are today a number of other drug treatments which many women are understandably reluctant to take. The team at Belfast divided 350 pregnant women into three groups: one in which patients were given acupressure to a point that Chinese physicians believe effective (a point near the right elbow called 'P6'); one in which patients were given acupressure to a part of the body not thought to have any effect, and a control group who received no treatment.

All patients were questioned regularly on a number of consecutive days about their sickness symptoms and rated their severity. The finding was that while the patients receiving dummy treatment experienced no relief, as did the control group, the group receiving acupressure to the P6 point experienced 'significantly less severe symptoms'.[10] The same team had carried out and reported three other trials published in the *Lancet*, *British Medical Journal* and *British Journal of Clinical Pharmacology* which were also positive. Here they had convincingly demonstrated the usefulness of acupressure and acupuncture in reducing post-operative sickness and nausea associated with cancer chemotherapy.[11] A separate acupressure trial has also reported successful use of the technique to reduce post-operative vomiting.[12]

In 1990, the *Lancet* reported a major trial in the United States to determine whether changes of lifestyle can reverse coronary heart disease.[13] In the trials twenty-eight heart patients were prescribed changes in lifestyle and compared with a control group of twenty who were treated by conventional methods such as cholesterol-lowering drugs and bypass surgery. The kind of lifestyle changes prescribed included a low-fat vegetarian diet, stopping smoking, training in how to manage stress, and moderate exercise.

One big question the trial set out to answer – quite apart from the effectiveness of the treatment itself – was whether outpatients could be motivated to change their lifestyle so much for long periods. The study showed both that such change is possible and that its results are similar to those gained using drugs. 'The changes in serum lipid levels are similar to those seen with cholesterol lowering drugs. The lifestyle intervention seems safe and compatible with other treatments of coronary heart disease.' Moreover, 'After a year, patients in the experimental group showed significant overall regression of [furring up of the arteries].'

In 1985, the journal *Holistic Medicine* carried an article by Dr Ashley Conway, a Harley Street psychologist who has used hypnotherapy effectively to treat a number of disorders. Conway's paper reviewed the research relating to the use of hypnotherapy in cancer treatment. Conway, too, is cautious in his assessment and says 'Research into the uses of hypnotherapy in cancer treatment is fraught with difficulties of both a practical and ethical nature.'[14]

Nevertheless, Conway concludes that, 'While claims that morbidity may be directly influenced by behavioural means are likely to remain controversial, it appears likely that with regard to symptom control some form of psychological intervention may benefit the cancer patient.'

In another article Conway also suggested that some major areas of medical research are being ignored completely because they do not fit in with the mechanistic reductionist paradigm adhered to by orthodox medical research institutions.[15]

> The mind–body link provides an enormous field for innovative research, and so far we have not even scratched the surface. In cancer chemotherapy anti-emetic medication is often not effective in controlling nausea. Hypnosis is. Why? 'Spontaneous' is an adjective frequently put in front of the word 'remission' by doctors, with the

implication that where remission does occur it is by chance, a statistical freak. A more honest expression might be 'unexplained remission' because there is mounting evidence to suggest that remission occurs in people with certain types of personality and belief systems.[16,17] Could counselling or psychotherapy improve a patient's chances of experiencing a 'spontaneous' remission? How does the placebo effect work and why are we not consciously exploiting it in a positive way? It is clear that major emotional trauma is accompanied by immune deficiency and that morbidity and mortality increase following bereavement.[18,19,20] Why do some people suffer more than others? Again could counselling or psychotherapy make a difference? Immunosuppression can be behaviourally conditioned and Ghanta *et al.* have demonstrated conditioned immuno-enhancement.[21] What are the implications of such work? The Simontons' work suggests that visualisation can influence physiology. Is this the case, and if so, what are the limits?[22] Anxiety has been linked with subsequent major . . . heart disease.[23] Could psychological techniques to reduce anxiety reduce the incidence of morbidity and mortality?[24]

If Conway is right in only one tiny part of this shopping list of research topics, the resulting treatments or preventions could save thousands of lives each year: lives that are currently being lost. Yet despite the wealth of evidence that now exists connecting lifestyle and psychological factors with disease, it is almost impossible for any researchers in these areas to obtain funding to conduct further research, either from government or from the major private medical charities.

The reluctance of funding bodies to back this kind of research stems from two sources: first because it is concerned with taboo areas, and second because it is concerned primarily with prevention rather than cure. Ever since antibiotics revolutionised medicine in the 1940s by providing a *scientific* drug-based cure for a wide range of medical problems, medicine has been primarily concerned to discover, through scientific research, further 'magic bullet' cures. This idea is the primary driving force behind all officially funded cancer research and behind much research into cures for heart disease, stroke and Alzheimer's disease – together the four biggest killers.

The Medical Research Council, which governs publicly funded

medical research, and private charities such as the Imperial Cancer Research Fund, are interested only in funding research that will discover scientific cures, preferably drug-based cures. They are not interested in techniques such as stress counselling or relaxation through meditation which have been shown to be effective in preventing and reversing disease. Their single-mindedness springs from a number of sources. First, much of their funds are derived from the major drug companies. These firms contribute cash directly (the government estimates their direct contribution to academic and hospital research currently as in the range of £70 million to £100 million) and also make important beneficial indirect contributions in areas such as sponsoring conferences, and the financial support of approved medical publications through extensive advertising.[25] Not unnaturally, they are able to influence research in the direction of being drug related. For such drug companies, there are no millions in meditation.

But the single-minded devotion to 'magic bullet' cures springs also from medical science's dazzling succession of cures in the past 100 years: vitamins, insulin, penicillin, cortisone, and drugs like Valium and beta-blockers. No medical problem is apparently too tough for science to crack if enough resources are put into it, with the result that many billions are poured into such research each year, while those seeking modest sums for holistic treatments find funding increasingly difficult to get.

American heart disease researcher Dean Ornish was responsible for developing the successful regime of diet and relaxation for cardiac patients described earlier. But he was unable to get funds from the American government or the American Heart Association. 'They said it was impossible to reverse heart disease,' explains Ornish. 'They said you need to use drugs because you can't motivate people to change their ways over a long time.' In the end, Ornish found his financial backing from successful businessmen in oil and real estate. 'Medical training is funded by drug companies,' he says. 'So are medical journals and scientific meetings.'[26]

It is a sobering thought that in this medical drug-oriented society more than half of all adults take some kind of drug every day; that 30 per cent of all hospital patients suffer unwanted side-effects from drugs; and that one in seven of all hospital beds in the United States is occupied by patients under treatment for adverse reactions to drugs.[27]

But if the nation's health is at stake, and both government and private funding bodies are taking a blinkered approach to research, we must inevitably ask just how far are Britain's scientific research policies rationally directed and how far are they affected by taboo? How effective are our current scientific policies? And should we be considering changing them?

CHAPTER SEVEN

Forbidden Fields

*To make our idea of morality centre on
forbidden acts is to defile the imagination.*
ROBERT LOUIS STEVENSON
Across the Plains

The taboo reaction in science that has been exemplified in previous chapters takes many distinct forms, all closely related but differentiated by their effects. At its simplest and most direct, tabooism is manifested as derision and rejection by scientists (and non-scientists) of those new discoveries or new inventions that cannot be fitted into the existing framework of knowledge. The reaction is not merely a negative dismissal or refusal to believe; it is strong enough to cause positive actions to be taken by leading sceptics to compel a more widespread adoption in the community of the rejection and disbelief, the whipping up of opposition, and the putting down of anyone unwise enough to step out of line by publicly embracing taboo ideas.

The simple taboo reaction is seen in the rejection by the Académie des Sciences of belief in meteorites and the shaming of former believers into throwing away specimens that had been collected; the rejection of the Wright brothers' claim to have flown, with 'proofs' offered by scientists of the impossibility of powered flight to guide the uninformed; the rejection of Edison's electric lamp with public derision by professional scientists of his amateur status and the scientific impossibility of his claims. The taboo reaction in such simple cases is eventually dispelled because the facts – and the value of the discoveries concerned – prove to be stronger than the taboo belief, as in the case of cold fusion. But there remains the worrying possibility that many such taboos prove stronger (or more valuable) than the discoveries to which they are applied – Tesla's turbine, or the effectiveness of holistic treatment of cancer and heart disease, for example.

In its more subtle form, the taboo reaction draws a circle around a subject and places it 'out of bounds' to any form of rational analysis or investigation. In doing so, science often puts up what appears to be a well-considered, fundamental objection, which on closer analysis turns out to be no more than the unreflecting prejudices of a maiden aunt who feels uncomfortable with the idea of mixed bathing. This form of scientific taboo is best seen in the prohibition against investigating any form of electromagnetic field associated with living organisms, when there is actually very substantial physical evidence for such a field.

The penalty associated with this form of tabooism is that whole areas of scientific investigation, some of which may well hold important discoveries, remain permanently fenced off and any benefits they may contain are denied us. In the example given, this could be of exceptional importance because the existence of biological electrical fields may enable us to diagnose illnesses, including viral infections and cancer, at a far earlier date.

Subtler still is the taboo whereby scientists in certain fields erect a general prohibition against speaking or writing on the subjects which they consider their own property and where any reference, especially by an outsider, will draw a rapid hostile response. This is clearly seen in fusion research where the incursion by scientists from another field – electro-chemistry – provoked public charges of incompetence, delusion and even fraud. But it is also seen in other important areas of science where a non-scientist presumes to speak or write about a topic of general interest; such as evolutionary biology.

Once again, the cost of such tabooism is measured in lost opportunities for discovery. Any contribution to knowledge in terms of rational analysis, or resulting from the different perspective of those outside the field in question, is lost to the community.

In its most extreme form, scientific tabooism closely resembles the behaviour of a priestly caste who perceive themselves to be the holy guardians of the sacred creed, the beliefs that are the object of the community's worship. Such guardians feel themselves justified by their religious calling and long training in adopting any measures to repel and to discredit any member of the community who profanes the sacred places, words or rituals, like the medical scientists who refuse to countenance acupuncture or hypnotherapy as anti-nausea treatments for their cancer patients.

Sometimes, scientists who declare a taboo will insist that only they are qualified to discuss and reach conclusions on the matters that they have made their own property; that only they are privy to the immense body of knowledge and subtlety of argument necessary fully to understand the complexities of the subject and to reach the 'right' conclusion. Outsiders, on the other hand, (especially non-scientists) are ill-informed, unable to think rationally or analytically, prone to mystical or crank ideas and are not privy to subtleties of analysis and inflections of argument that insiders have devoted long painful years to acquiring.

Yet the history of science abounds with examples that contradict this kind of elitist thinking: amateurs or non-scientists like the Wrights, Edison and even Charles Darwin; and outsiders like John Dalton (the self-taught meteorologist who revolutionised chemistry) and Fleischmann and Pons, the chemists who 'rescued' fusion physics.

Perhaps the most worrying aspect of the taboo reaction is that it tends to have a cumulative and permanent discriminatory effect: any idea that is ideologically suspect or counter to the current paradigm is permanently dismissed, and the very fact of its rejection forms the basis of its rejection on all future occasions. It is a little like the court of appeal rejecting the convicted man's plea of innocence on the grounds that he must be guilty, or why else is he in jail? And why else did the police arrest him in the first place? This 'erring on the side of caution' means that in the long term the intellectual Devil's Island where convicted concepts are sent becomes more and more crowded with taboo ideas, all denied to us, and with no possibility of reprieve.

How exactly do the guardians of the temple of knowledge impose their taboo? On the face of it, even the suggestion is implausible. Surely scientists, however prejudiced some individuals may happen to be, lack a mechanism to impose their world view on others in any systematic way? In practice, it is the easiest thing in the world for those who manage the profession of science to declare a subject taboo. For any young scientist to progress in the profession, or even to be employed in science at all, he or she has to conduct research (or to teach what has already been learned from research to students). But to conduct research, the scientist must receive funds from some institution or organisation and getting such funds means placing proposals before a number of scientific colleagues who will evaluate them and decide whether or not to recommend that the financial and other necessary resources should be used in this way.

At this stage, the research scientist may be persuasive enough to secure funding or the institution enlightened enough to grant the resources, even for research which apparently flies in the face of orthodox scientific belief (although this happens less and less often today). But the researcher will then encounter the next obstacle which is usually insuperable. To continue to receive funds, he or she must publish their initial findings. A paper has to be submitted to one of a few professional journals where it will be reviewed by distinguished senior scientists from the field. Nominally, these reviewers are the researcher's scientific peers. In reality they are likely to be academically superior. Even if not senior in rank they are superior in at least one important respect: they decide who gets published and who doesn't. If they believe the research to be without merit (regardless of the results obtained) they will block its publication. And unless the researcher can get some findings published, he or she will find it impossible to get the grant renewed. Unless there are very unusual extenuating circumstances, non-publication is taken in science to mean experimental failure.

The popular weekly magazine *New Scientist* said in an editorial column:

> The process of getting a paper into print has evolved to match the scientific environment of the day. . . . Nowhere in this process is there any talk of publicity or other base motives, but in a world where researchers fight for every penny, not to mention their jobs, 'PR' is an inevitable factor. Some of the more influential scientific journals know this only too well. Thus some editors warn authors that if there is so much as a whiff of advance publicity for a forthcoming paper, the journal will withdraw it.[1]

Thus, on the one hand, there is pressure to achieve publication in approved journals, and at the same time pressure not to seek publicity elsewhere.

The net effect of this peer review system is that, at any given time, almost the entire research effort of the country is directed into subjects that are tacitly approved by those who comprise editorial review committees of the scientific press. Those review committees are, in turn, frequently drawn from among the more conservative scientists in the community and the system is self-perpetuating from supervisor to postgraduate to undergraduate. For this reason, virtually the only scientific research being conducted anywhere in Britain into taboo subjects is privately funded and is usually carried on in 'skunk works' – private laboratories with little or no resources – and its results are privately published usually in short-run paperback or photocopied editions that do not receive general circulation, or reach major libraries.

Is there any solid evidence to substantiate the charge that the senior members of the scientific profession discourage objective study of anomalous phenomena by ordinary working scientists? Perhaps surprisingly, there is evidence pointing to just such discrimination. The evidence comes from examining how scientists' attitudes to anomalous phenomena have changed over the past fifty years. In 1938, some 352 members of the American Psychological Association were surveyed on their attitude to the subject of extra-sensory perception – did they think ESP an established fact, or a likely possibility? As one might expect at a time when almost no serious research had been conducted into the subject, only 8 per cent of members thought ESP a legitimate subject for study, a figure likely to be very representative of the view of most scientists at that time. Forty years later, during which time hundreds of well-designed parapsychology experiments had been conducted, the opinion of scientists in general had changed considerably. In 1972, *New Scientist* surveyed its readership and found that 67 per cent thought ESP established or likely. And in 1979, a survey of more than 1,000 college professors in the United States found that as many as 75 per cent thought so.

However, in 1981, Dr James McClenon, a sociologist from Maryland University, studied the opinions of the 'administrative elite' of American science, the council and members of certain committees of the United States' most prestigious scientific body, the American Association for the Advancement of Science. What he found was that – while 75 per cent of college professors in general thought ESP an established fact and worth investigating – as few as 20 per cent of AAAS officers thought so. This very low score was found amongst the AAAS committee members from the social sciences – the discipline that would most likely be responsible for parapsychology research. Even amongst those scientific officials from disciplines such as physics and chemistry, the score was only 30 per cent, suggesting a very substantial gap in perception of the scientific value of the subject between those at the top and those on the laboratory bench.[2]

The effects of the peer review consensus are felt not merely in the closed world of scientific publications. Many book publishers (though, happily, not all) rely on the same groups of professional scientists that comprise the journal review panels to advise on the suitability of their textbooks and even publications for general consumption by the scientifically literate public. It is thus highly improbable that a taboo scientific subject will receive serious treatment, or wide circulation. As described in Chapter 11, when Macmillan stepped out of line by publishing a book that did not meet with the approval of the US scientific community, the company was compelled to cease publishing by an academic boycott of its textbooks.

Much the same considerations apply, for much the same reasons, to the broadcast media. Although nominally free to make programmes about any subject of legitimate interest, television producers would rarely make and broadcast a programme on a taboo science subject, because they naturally perceive it as more than their job is worth to offend the orthodox scientific community too often. Whenever a contentious subject is broadcast, the television company or channel is bombarded with complaints from professional scientists and for this reason some subjects are rarely attempted, and even then are treated in a softly-softly way.

Here are two recent examples. In 1990, television film maker Hilary Lawson, who has made a number of films under the banner of *Horizon*, *Equinox* and other well-known television series, made a programme on the subject of global warming. In the course of the programme the point of view of critics of the global warming theory was explored and the evidence against the idea examined. The evidence presented showed that a number of beliefs held by those scientists who support the idea of global warming may be incorrect.

When the programme, entitled *The Greenhouse Conspiracy*, was broadcast on Channel Four in the UK in 1990 it attracted widespread criticism and condemnation as irresponsible from some orthodox scientists. When the programme was bought by the Public Service Broadcast network in the United States and was viewed by the editorial board, they decided to reject it for broadcast. The decision became a *cause célèbre* since and rejection of the programme was seen by many as scientific censorship. The issue was debated in Congress as being an unconstitutional suppression of free speech, a view which Congress upheld. It directed the PSB to broadcast the film – a direction which the Board obeyed in the letter but evaded in spirit by broadcasting it only in federal territory – Washington DC. The majority of the American public has thus been denied an opportunity to hear the purely scientific issues of global warming seriously debated in a public forum.

There is a temptation to imagine that such things could never happen in England, home of free speech, tolerance and civilised parliamentary debate. The true position is uncomfortably different from this traditional picture. Although the contentious global warming programme *was* publicly broadcast here, and not suppressed, there are many other taboo subjects that would never get made into programmes in the first place because producers and film makers know they would never get funding, and if they did, the resulting trouble would be bad for their careers. In addition, Congress was able to debate the issue of suppression of free speech only because it is guaranteed by the American constitution. In this country, where free speech is merely a tradition and not a legal right, Parliament would be unable to take up such an issue. Thus censorship in Britain tends to be not a public issue but a private matter: it is self-censorship and it is settled behind the scenes, not in public.

Another example is provided by the subject of evolutionary biology. In this country only the Darwinian mechanism of random mutation coupled with natural selection may be publicly discussed or taught in schools even though there is strong direct evidence from botany and microbiology of the inheritance of acquired characteristics under certain circumstances and of other evolutionary mechanisms more important than natural selection.[3] No science programme has ever been shown or is ever likely to be shown on British television questioning any aspect of the Darwinian mechanism, because it is the most strictly observed taboo subject.

Censorship of anti-Darwinian ideas is rather more overt in the United States. In California, for example, state authorities are trying to close down a religious graduate school offering science courses that include scientific creationism and other anti-Darwinian ideas on its syllabus. The graduate school was founded in 1981 by the California-based Institute for Creation Research, a religious organisation. It was licensed to teach by the state for a decade but in 1990, Bill Honig, superintendent of public instruction, appointed a five-strong team of anti-creationist scientists to investigate the graduate school. The team recommended that the state deny a licence to operate to the school on the grounds that it is teaching anti-Darwinian ideas at graduate level.

In sharp contrast, in Moscow in May 1990, the issue of evolution itself was debated in prime time on Soviet National Television's educational Channel 1, which broadcasts over the entire Eurasian land mass from Finland to Japan. The debate was sponsored by and held under the aegis of the Soviet National Academy of Science. Speaking against Darwinism were Dr Leonid Korochkin of the Academy's Institute of Developmental Biology together with American scientific creationist Dr Duane Gish, while Darwinism was defended by Dr Georgi Gauze also of the Academy's Institute of Developmental Biology. It is deeply disturbing to learn that such public debate of a taboo subject can take place in a so recently totalitarian society, but is effectively forbidden in tolerant, liberal, democratic Britain and America.

Curiously, neither press nor television in Britain feels any inhibition about airing highly contentious political or social issues and will risk considerable controversy to assert their right to cover what they consider to be in the public interest. But when it comes to contentious scientific matters, they become much more reticent. Probably this is because giving space or air time to political controversy is good, for their image, whereas if they treat taboo science subjects too seriously, they risk being derided as mentally unbalanced or 'unreliable'.

Many scientists reading the description above of the state of science in Britain, might agree with the facts while disputing their interpretation. Yes, the peer review system is designed to be a tough test which research results must pass or fall by the wayside; and yes, science jealously guards the keys to its kingdom, for the world really is full of cranks and obsessives with half-baked ideas, mainly long ago discarded, who would waste valuable public resources unless we take care. Yes, we must guard the current paradigm against assault by the ignorant and the mad masquerading as the knowledgeable, or the whole of our knowledge will become debased and valueless, we will return to the dark age when mankind was subject to the tyranny of opinion.

There is no question that there are important truths in these arguments, truths that we dare not ignore. It would be criminally negligent to waste public research funds on studying ideas already known to be valueless and this is one of the most important services that scientists can perform for us in allocating the community's scarce resources. But while taking these important points into account we must look at the large picture and question just how often the present system delivers the goods in terms of research and how often it is those whom the system rejects that do the delivering.

Fortunately this is not an issue on which we have to take anybody's word, however persuasive their arguments, for it is a matter of public record. How often in the past 100 years has publicly funded institutional scientific research resulted in major scientific and technical innovations? And how often have such discoveries come from the derided and ostracised loner in his 'skunk works'? Even a superficial review shows a massive trend in favour of the latter.

Bell and the telephone; Parsons, Tesla and the turbine; Edison and the electric light and recorded sound; Marconi, Tesla and radio; the Wright Brothers and flight; Carl Benz and the automobile; the Lumière brothers and cinema; Otto Mergenthaler and the Linotype machine; Armon Strowger and the automatic telephone exchange; George Eastman and celluloid photographic film; Fritz Haber, the historian who taught himself chemistry and fixed atmospheric nitrogen; Wegener and continental drift; Pollen and automatic fire control; Baird and television; Whittle and the jet engine; Chester Carlson and Xerography; Eckert and Mauchly and the commercial computer; Edwin Land and Polaroid Photography; Christopher Cockerell and the hovercraft. Even where the innovator belongs to a recognised institution he or she is often a loner who achieves success by swimming against the currents of orthodoxy, like Alan Turing and the first British computers, or even Watson and Crick, who had been told to drop their study of DNA but continued it as 'bootleg' research.

Of course, one can also compile a long and distinguished list of discoveries in institutional science, especially from the great universities like Oxford and Cambridge and especially in important basic fields such as atomic physics and astronomy. But, somehow, it is difficult to draw up a list that carries quite the same diversity, the same romantic air of excitement and innovation and one that has so obviously influenced every single aspect of twentieth-century life so fundamentally. Anyone who switches on the electric light, turns on the television, makes a phone call, watches a film, plays a record, takes a photograph, uses a personal computer, drives a car or travels by aeroplane has the lone eccentric to thank, not institutional science. And important though the nation's programme of scientific research is, it seems to me that this review should be profoundly disturbing to anyone who cares about how efficiently public money is spent on scientific research.

Whatever else it might be guilty of, no one can accuse the government of failing to take science seriously. In Britain in 1990, a total of £12,137 million was spent on scientific research, of which 50 per cent was provided by industry and about 14 per cent by the generosity of private endowments. The remaining 36 per cent, some £4,369 million, was provided by central government. There has been constant growth in expenditure in real terms in recent years, averaging about 12 per cent, and plans for the immediate future are for even more growth (the 1993–4 budget represents a 14 per cent increase in real terms over 1990–1, for example).[4]

In 1991–92, total net government spending on research and development, for both civil and defence projects, amounted to £5,074 million, of which £2,824 million was spent on civil science in areas such as engineering, medicine, agriculture and food.[5]

To put this in perspective, spending on scientific research in 1992–3 amounted to more than 2 per cent of total expenditure and hence was a big budget item. It was more than the government spent individually on the departments of Agriculture, Trade and Industry, Employment, or Environment (apart from housing); more even than spent by the Foreign and Commonwealth Office. It was twice as much as was spent on the whole of Wales and equivalent to the total sum spent on Scotland.[6]

Publicly funded research is administered by five research councils: the Science and Engineering Research Council (SERC), the Medical Research Council, the Natural Environment Research Council, the Agricultural and Food Research Council, and the Economic and Social Research Council. In addition, expenditure on science together with university funding is done through the Universities Funding Council. Together these organisations cover an almost unimaginably wide and diverse range of projects. An idea of what the money is spent on can be gained from looking at SERC. It maintains four research establishments: the Rutherford-Appleton Laboratory in Oxfordshire, the Daresbury Laboratory at Warrington, the Royal Greenwich Observatory at Cambridge and the Royal Observatory at Edinburgh. In 1990–1, the council spent £130 million on science, £124 million on engineering, £87 million on nuclear research, and £77 million on astronomy.

Other big spenders in research are the Universities Funding Council who spent £930 million in 1990–1, and the Ministry of Defence, who spent £2,190 million on R and D.

Who oversees this enormous expenditure of public money? In principle, at the highest level, there are three organisations with a significant influence over research expenditure: the Public Accounts Committee (PAC), the House of Lords Committee on Science and Technology, and the Office of Science and Technology. The taxpayer's interest is represented by the Public Accounts Committee, an all-party body in Parliament with powers to summon and question both ministers and individuals. The PAC is certainly no tame rubber-stamp committee and it has cut up rough on a number of occasions in recent years when its members have felt the public was not getting value for money – for example, in the case of the health service. The Lords Committee similarly has shown that it does not tamely follow the conventional view and (as reported later) has championed the National Health Service as a research organisation. The Office of Science and Technology is at the heart of government, in the Cabinet Office, acting on behalf of the prime minister. Though relatively new, it, too, has a significant role to play in directing the nation's resources, and especially in coordinating research across more than one government department.

However, *none* of these organisations has ever seriously questioned expenditure on scientific research, believing as most administrators do that technical questions should be left to scientific experts and that the unpredictable nature of research means that scientists must be given the maximum possible leeway to conduct their chancy business. Instead of attempting to impose ill-informed dictatorial policies from outside, politicians have been content to allow scientists themselves, in the shape of the various science research councils, to administer the allocation of research funds.

In one important sense, Parliament's wishes are being carried out as intended, in that careful consideration is given by the research councils to the allocation of their limited resources, and scientists who wish to make claims on those resources are compelled to compete for grants in the most devoutly Darwinian manner. It is thus to be expected that only the fittest research projects will survive the fierce competition of this scrutiny to make claims on the public purse.

But there is another important sense in which Parliament's wishes are not being observed and in which, I believe, the taxpayer's interests are not being properly served. In essence it is that, though the individuals who sit on the various science research councils are undoubtedly people of first-rate intellect and the highest integrity, they are subject to a systematic bias towards conservatism that is built into the very fabric of scientific research itself; and, moreover, a bias that works *against* the most important part of the discovery process.

This bias shows itself in a variety of ways. At its simplest it is revealed in the peer review system described earlier. It shows itself in the tendency to proscribe certain areas of research as being untouchable. It also shows itself in some even more subtle ways in institutional research that are the subject of the second part of this book.

Above all, however, the current system of allocation of research funds means that there are virtually no circumstances where anyone – whether scientist or lay person – can seriously call into question the validity of any project, or any line of research. Specifically, in the British scientific research system as it is presently structured, and unlike almost every other area of public expenditure, there is no mechanism for questioning whether we, the taxpayers, are receiving value for money for the scientific service that we pay for.

Although it is impossible to examine any particular research project and to predict with any confidence what, if anything, will come of it, it is by no means impossible to evaluate critically the historical performance of publicly funded science over recent years, and to ask some key questions. For example, how has one of the biggest areas of expenditure, medical research, contributed to the nation's health?

The answer may come as a shock to anyone who takes it for granted that the billions of pounds we have spent on medical research have made a big difference to the incidence of disease and death. First, *life expectancy at age forty-five has barely changed since the beginning of this century*.[7] The statistical increase in life expectancy overall during that period has come about because there used to be a high incidence of disease and death in babies and children which has been eliminated by measures such as smaller families, better hygiene, birth in hospital instead of at home, antibiotics and health education.

Most of these preventive factors were in place and fully effective thirty or forty years ago. Their effect has been such that *life expectancy at any age has not changed significantly in the past twenty years*.[8] Yet, in that twenty-year period, Britain has spent around £2,000 million on medical research (at 1993 prices).

Turning to specific areas of medical research, what progress has been made in combating our biggest killers, such as cancer and heart disease? Once again the answers are deeply disturbing. Dr David Horrobin, then director of the Efamol Research Institute in Canada, writing in the *New Scientist* in 1982, said:

> Lay organisations whether charities or governments, do not fund

medical research for the sake of culture. They provide money because they believe that practical benefits will follow. It is gradually dawning on the donors that for the past 20 years practical benefits have not followed. During that time there have been no substantial improvements in morbidity or mortality from major diseases that can be attributed to public funding of medical research. The much vaunted successes in some relatively rare cancers, such as Hodgkin's disease, derive from refinements of discoveries that were made in the mid-1950s. The only substantial advances have come from the pharmaceutical industry. Even there, the foundation is so shaky that the two biggest successes, the beta-blockers (now widely used in treating hypertension) and the histamine-2 antagonists (effective against stomach ulcers), had to come from the mind of one man, Jim Black.[9]

Horrobin is not alone in his assessment and neither have things changed materially in the decade since he wrote these words. Take the fight against cancer, for example. In the *New England Journal of Medicine* in 1986, Dr John Bailar of the Harvard Medical School and Dr Elaine Smith of the University of Iowa Medical Center asked what progress we have made against cancer in the three decades from 1950 to 1982. Their conclusion was that:

> We are losing the war against cancer, notwithstanding progress against several uncommon forms of the disease, improvements in palliation, and extension of the productive years of life. A shift in research emphasis, from research on treatment to research on prevention, seems necessary if substantial progress against cancer is to be forthcoming.[10]

The two researchers found that in the United States from 1950 to 1982 the overall rate of death due to cancer had actually *increased* by 8.5 per cent and this during a period of unprecedented expenditure on research and clinical investigation.

These findings, of course, contrast sharply with the picture painted by governments and by some research funding agencies, and the results were hotly disputed. But the US General Accounting Office, in April 1987, published figures that confirmed there had been little or no improvement in patient survival for the twelve most common cancers from 1950 to 1982, despite expenditure of $1,000 million a year on research.[11]

The picture is no different in the United Kingdom. There has been some welcome progress in understanding how cancer attacks the body and a few of the more uncommon forms have shown dramatic evidence of improved treatment, such as leukaemias and non-Hodgkin's lymphoma. But there has been almost no change in very common and widespread forms of the disease, such as cancer of the breast and lungs in women, or prostate and bladder in men. At the same time, the incidence of cancer has increased considerably, offsetting the gains that have been made. Amongst British men, for instance, there were 2,721 deaths per million from cancers of all types between 1971 and 1975: compared with 2,970 deaths per million between 1981 and 1985, an increase of 9 per cent. For women, the increase over the same period was even higher, at 14 per cent.[12]

In the February 1991 issue of *Journal of the Royal Society of Medicine*, Dennis Burkitt, who discovered the link between cancer and lack of dietary fibre, wrote that, 'Little real success has been achieved,' and that 'most cancer research is misdirected.' Burkitt called for a new strategy focusing on prevention because, 'medicine has waged a major war against cancer, concentrating on earlier diagnosis and improved therapy. The war is not being won.'

Dr Robert Sharpe, writing in the Animal Aid journal *Outrage*, said that:

> Fortunately, with the recent revival of interest in epidemiology, doctors know far more about the real causes of cancer, so that 80–90 per cent are considered potentially preventable. Nevertheless the government and cancer research organisations seem reluctant to face the challenge of preventive medicine. In 1979, the *British Medical Journal* criticised the Cancer Research Campaign (CRC) and Imperial Cancer Research Fund (ICRF) for paying so little attention to prevention and noted that the CRC spent less than 2 per cent of its funds on cancer education. Ten years later, the CRC allocated just 1.5 per cent of its £35 million expenditure to cancer education, whilst the ICRF, with an income of over £47 million, does not even mention an educational budget in its annual accounts.[13]

As evidence of the lack of will on the part of government to confront preventive medicine, Sharpe points out that there is overwhelming evidence that both alcohol and tobacco are major cancer-causing agents (tobacco was responsible for a more than 40 per cent increase in deaths through lung cancer in women from 1975 to 1985) yet the government's response is weak and ambiguous. When the chairman of the British Medical Association called in 1993 for a complete ban on tobacco advertising on health grounds, the Health Secretary, Virginia Bottomley, would say only that the matter will be 'kept under review'.

One further aspect of UK medical research that many people will find surprising is that, after forty-five years of existence, the National Health Service plays almost no part in medical research. The House of Lords Select Committee on Science and Technology, commenting in 1988 on the widespread despondency in the NHS, said that 'The overriding cause of the collapse of morale is the impression that neither the NHS or the Department of Health and Social Security demonstrates any awareness of the importance of research nor is prepared to devote time, effort, resources, to promote it.'

The NHS provides many examples of excellence, and when it has been given resources to conduct research it has produced some stunning results. They include *in vitro* fertilisation or test-tube birth, developed by NHS gynaecologist Patrick Steptoe, Cambridge scientist Robert Edwards and their chief scientific assistant Jean Purdy; the artificial hip joint whose development was contributed to mainly by John Charnley at Manchester Royal Infirmary; and computer-aided diagnosis of abdominal pain developed by Tim de Dombal and his team at St James's University Hospital Leeds.[14]

There is a widespread and growing belief amongst medical scientists that there should be a switch in emphasis from treatment to prevention. That is the main contention of the British Holistic Medical Association and of supporters of the National Health Service as a research organisation. It is certainly easy to follow the arguments in favour of this approach, primarily that it is in treating real patients in general practice and hospitals that the greatest opportunities present themselves for effective research and for devising and testing preventative measures.

One outstanding example is that of GP Norman Beale, reported in

New Scientist.[15] Dr Beale investigated the closure of a factory in Calne, Wiltshire, to find out if there is a link between unemployment and poor health. In a statistical trial, he found that unemployment increases consultation rate by 20 per cent and referral rates to out-patients departments by 60 per cent. These referrals are due mainly to chronic illnesses such as heart disease and high blood pressure, which are six times greater among the out of work than amongst employed people.

Reviewing the evidence of the last two chapters it seems to me that there is a strong case to be made for some fundamental changes to our national policies on scientific research. The following is a list of suggestions for discussion:

End-user involvement

Scientific research is the last great area of public expenditure where the end user has no voice at all. This failure must be remedied, not only as a matter of social justice but also because it will strengthen the direction and quality of research, especially in key areas such as medicine.

By way of example, there is strong experimental evidence of at least three ways in which morning sickness in pregnancy can be reduced and prevented (acupressure, acupuncture and hypnotherapy) as described in Chapter 6, but no public money is devoted to large-scale trials of these techniques and none of the treatments is obtainable through the NHS. When you consider that there are about three-quarters of a million pregnancies in Britain every year and that *up to 600,000 of those women* will experience morning sickness ranging from mild to severe, I believe that this lack of research and treatment is a national scandal. I cannot help wondering how far the composition of the Medical Research Council influences its decisions on issues of this kind: in a recent list of its twenty-five members, only two were women. If the MRC paid attention to the voice of end users of scientific research, however informal, then I feel certain research funds would already have been made available for such a beneficial purpose. Moreover, if the incidence and severity of morning sickness were significantly reduced, many employers would cease to perceive pregnancy as an illness, as they do at present.

There is clearly a strong case for user councils in scientific research just as there are in the provision of other key public services, such as public utilities and transport.

Positive discrimination

As things stand, the taboo reaction in science tends to have a cumulative one-way effect. As described earlier, once a discovery has been condemned, it tends to remain condemned – the Devil's Island syndrome. This means that research institutions are being permanently deprived of important raw material for investigation; that the flow of ideas is itself being stifled; and that there is no appeal against the discrimination. In the long term, this may well mean that not only is the community being deprived of beneficial discoveries, but that the omissions can never be rectified.

To combat this tendency to permanent exclusion I propose two things. First, that a small scientific team be appointed to examine critically the corpus of taboo ideas; it reconsider the existing published research in the light of Sturrock's criteria (see Chapter 11), including the deliberate exclusion of probability values of zero (complete disbelief) and unity (complete certainty). Second, I suggest that the same team have the power to read and, at their discretion, listen to presentations from researchers who believe they have been dismissed for non-scientific reasons. If they agree, they should also have the power to disseminate the research paper to libraries, information departments and research laboratories, drawing their attention to its possible usefulness. They would thus serve as scientific ombudsmen.

The team I envisage will have to be interdisciplinary and have access to the services of specialists, cover a very wide range of interests, be exceptionally open-minded to new ideas – and also have endless patience, since they will inevitably become a target for every obsessive and crank in the country. Many scientists will believe that this alone is reason enough not to adopt such a policy. I believe it is the very reason for doing so. It is time to start listening: we have ignored people long enough.

Prevention is better than cure

The evidence for effect of lifestyle on some major illnesses such as heart disease and cancer shows a clear case for nationally funded and directed research aimed at formulating large-scale trials of preventative measures.

National Health Service involvement

The finding of Norman Beale's study linking unemployment and disease is the kind that is of fundamental importance to medical research, but it could only ever come from the real world of runny noses and the dole queue. It could never emerge from the clinically clean laboratories of pure research. The NHS has an essential role to play in developing preventative measures of the sort mentioned above.

I also propose that every committee sitting to discuss the expenditure of medical research funds, especially those of the Medical Research Council, be obliged to include working general practitioners and hospital doctors whose primary daily task is getting people better, and that the needs of getting patients better should generally take precedence over all other considerations.

Open door policy

Under the general heading of a policy designed to make use of, rather than exclude, taboo ideas that might perhaps prove useful, and which could be carried out by the team mentioned above, the following ideas might also be worth considering:

- The setting up of some form of central collation office, possibly based on a computer database, whose task will be to seek correspondences between taboo areas of research and make suggestions for consideration by the research councils.
- Some form of active collaboration with the lone researchers in their 'skunk works' – even if only at the level of exchanging ideas.
- A one-off trawl through taboo research and results to see if anything can be used or fitted into current research.

The first part of this book has been concerned primarily with analysing *how* institutional science has marginalised certain contentious areas of research that orthodox scientists perceive as 'dangerous' or cranky and the cost to the community of this taboo reaction. The second part is concerned with the rather more difficult questions of just *why* some scientists are motivated to behave in this way; what the effects of their behaviour on the development of science within the community are; and how we can tell a real crank from a researcher who merely stumbles accidentally across a subject that is taboo to orthodox science. To even begin to get a grasp on this difficult subject we must first pay a visit to the psychology laboratory to examine much more closely the question of how we perceive natural phenomena.

PART TWO

Taboo

Science is not powerful because it is true.
It is true because it is powerful.

HILARY LAWSON

CHAPTER EIGHT

Calling a Spade a Spade

When I see a spade, I call it a spade.
OSCAR WILDE
The Importance of being Earnest

When Cecily Cardew confronted her rival for the hand of Ernest Worthing in Wilde's famous duel over the tea cups, she declared that when she sees a spade, she calls it a spade. Cecily might have been less uncompromising and confident had she been a subscriber to the *Journal of Personality* and read the paper by J.S. Bruner and Leo Postman, published in 1949.[1]

The two psychologists wanted to discover exactly how people perceive and understand visual symbols. To help them, they designed and made some special packs of playing cards. Some cards were quite normal, but others were altered in subtle ways; for instance the six of spades was turned into a red card and the four of hearts was made black. The anomalous cards were mixed with normal cards, the pack was shuffled and test subjects were shown the cards one at a time.

At first, people were shown the test card for merely a short glimpse. Gradually they were shown each test card for a longer and longer time until they were able to recognise and identify it to the researchers. Even with only a short glimpse, most people were able to identify all the cards shown. But the extraordinary finding was that the anomalous cards were always identified as being normal without any hesitation or puzzlement.

The people looking at them actually saw the black four of hearts as either a four of spades or as a normal four of hearts. Their perceptions were simply fitted naturally into the conceptual categories that had already been prepared by their previous experience of playing cards.

When the experimenters increased the amount of exposure to each anomalous card, people began to become aware of something wrong. One subject shown the red six of spades said: 'that's the six of spades but there's something wrong with it – the black has a red border'.

Further increases in exposure time made subjects even more confused and hesitant, until most people finally 'saw' what was really before their eyes. Most interesting of all, however, was that more than 10 per cent of the anomalous playing cards were *never* correctly identified even when exposed for forty times the average exposure needed to recognise normal cards. And many of the people taking part in the test experienced acute personal distress. One person remarked to the experimenters, 'I can't make the suit out, whatever it is. It didn't even look like a card that time. I don't know what colour it is now or whether it's a spade or a heart. I'm not even sure what a spade looks like. My God!'

Even the experimenters themselves, who knew every card in the phoney deck, were disturbed by viewing them. Postman told a colleague that, 'though knowing all about the apparatus and display in advance, he nevertheless found looking at the incongruous cards acutely uncomfortable.'[2]

This mental discomfort, and our attempt to avoid it, extends beyond perception of the mere physical symbols themselves and embraces the meaning or significance of those symbols. At around the time that Bruner and Postman were asking questions about how we perceive things, Leon Festinger and his colleagues at Stanford University were formulating a theory about how we *believe* things and how those beliefs affect our behaviour. Festinger proposed the theory of cognitive dissonance – that we all strive to keep a sense of consistency between the things that we think we know and we resist any new information that causes dissonance between our beliefs, or we strive to reduce that dissonance.[3]

The kind of studies on which Festinger based his theory of cognitive dissonance seem rather obvious when looked at in a common-sense way. But they actually expose an important component of our thought processes (or, if you prefer, of our behavioural processes) that is normally invisible to us.

Take, for example, the survey poll carried out in Minnesota in which 585 people were asked, 'Do you think the relationship between cigarette smoking and lung cancer is proven or not proven?' The poll showed that the attitude of smokers and non-smokers to this question differed sharply. Among non-smokers, 29 per cent thought the link was proved and 55 per cent thought it not proved. Those who smoked heavily held very different views. Only 7 per cent of heavy smokers thought the link proved and a whopping 86 per cent thought it not proved.

The important question here is not the factual scientific question of who is right and who is wrong. It is why should smokers hold such a strongly different belief from non-smokers? The answer that Festinger gives is that the smokers are acting to reduce their level of cognitive dissonance by denying the link – despite considerable medical evidence. Knowing they smoke and accepting the medical evidence would create a distressing inconsistency in their beliefs. The simplest way to reduce that distress is to deny the new information.

Festinger generalised his theory to explain how people will tend to reduce cognitive dissonance that stems from social disagreement. The greater the magnitude of the dissonance – the more strenuous the efforts to reduce it. Festinger identified three mechanisms that we may use to try to reduce dissonance that stems from such disagreement.

The first and most obvious is to change our own opinion so that it corresponds more closely with our knowledge of what others believe. This explains the widespread phenomenon of the group viewpoint or the tendency of any group of people to wish to achieve a consensus viewpoint.

The second way is to try to apply pressure to those people who disagree to alter their opinion. This is an equally common phenomenon and one that explains just why some individuals are willing to go to such strenuous lengths to try to make others think as they think.

The third method, according to Festinger, is equally easily recognised:

> Another way of reducing dissonance between one's own opinion and the knowledge that someone else holds a different opinion is to make the other person, in some manner, not comparable to oneself. Such an allegation can take a number of forms. One can attribute different characteristics, experiences or motives to the other person

or one can even reject him and derogate him. Thus if some other person claims the grass is brown when I see it as green, the dissonance thus created can be effectively reduced if the characteristic of being colour-blind can be attributed to the other person. There would be no dissonance between knowing the grass is green and knowing that a colour-blind person asserted it was brown.

There is substantial experimental evidence to support this view. Schacter set up a complex series of experiments involving people brought together in 'clubs' to discuss how best to deal with young criminals. Without the knowledge of the test subjects, paid participants always adopted certain attitudes in the club debates that followed. One paid participant always agreed with the meeting, another always disagreed, saying for example that juvenile offenders should be harshly punished. The study found that people who persistently disagreed with the group's view were consistently derogated by the group and there was a move to exclude these people from future meetings of the 'club'. Even more interesting, half of the 'clubs' were made to seem very attractive to the participants, while the other half were made to seem considerably less attractive. The extent to which members derogated and wished to ostracise the person who disagreed with them was far higher in the attractive clubs than in the less attractive clubs.[4]

But, of course, not everyone reacts in the same way to learning new information that contradicts their existing beliefs. Festinger concluded that:

> For some people, dissonance is an extremely painful and intolerable thing, while there are others who seem to be able to tolerate a large amount of dissonance. This variation in 'tolerance for dissonance' would seem to be measurable at least in a rough way. Persons with low tolerance for dissonance should show more discomfort in the presence of dissonance and should manifest greater efforts to reduce dissonance than persons who have high tolerance.

> At this point many readers will feel like suggesting that perhaps such a test already exists, having recognised a certain similarity between our discussion immediately above and some descriptions of 'authoritarian personalities' and some descriptions of people with high 'intolerance for ambiguity'. My own suspicion would be that

existing tests such as the F scale do measure, to some extent, the degree to which people hold extreme opinions, that is, opinions where dissonance has been effectively eliminated.[5]

The authoritarian personality with low tolerance for dissonance and who readily adopts the device of derogating others is one that we have already met in previous chapters and will be meeting in a number of guises in later ones. The F scale referred to by Festinger is a measure of authoritarian tendencies, devised by American researchers to try to measure an individual's predisposition towards fascism. This is also examined in more detail later.

There have, of course, been many such basic findings about perception in experimental psychology over the past fifty years or more. The question is, what if anything do they show in the case of scientific discovery? One scientist who concluded that they show a great deal was Thomas Kuhn of Berkeley University, California, who originated the first comprehensive theory of how scientific revolutions come about.[6]

In his book *The Structure of Scientific Revolutions*, Kuhn popularised the now widely accepted idea of the scientific 'paradigm'; universally recognised scientific achievements that for a time provide model problems and solutions to a community of scientists engaged in those and related problems.

'In science,' says Kuhn, 'as in the playing card experiment, novelty emerges only with difficulty, manifested by resistance, against a background provided by expectation. Initially only the anticipated and usual are experienced even under circumstances where anomaly is later to be observed.'

This idea is one that many people, including scientists, will find simply impossible to accept. Are we really being asked to believe that when scientist 'A' looks at an experimental result he sees one thing, but when scientist 'B' looks at the same experiment he sees something quite different, because of differences in their personality? Extraordinary though it may sound, that is exactly the conclusion that Kuhn and others have reached. And the evidence from the history of science is not merely persuasive, it is overwhelming.

One of the most interesting examples that Kuhn cites is Sir William

Herschel's discovery of the planet Uranus – the first planet to be discovered since prehistoric times. This is interesting not merely because it shows the 'playing card' syndrome in action, but because it also triggered what Kuhn has called a paradigm shift in the branch of science concerned.

On at least seventeen occasions between the years 1690 and 1781, a number of astronomers, including some of Europe's most influential observers, had seen a 'star' in positions that we now know to have been that of Uranus. One astronomer had even observed the object for four nights in a row in 1769 but without noticing the motion that would have disclosed it as a planet not a star.

When Herschel first observed the same object twelve years later he was able to examine it with a much better telescope that he had himself designed and built. Herschel saw that the object appeared to have a disc shape – something not characteristic of stars because they are too far away to be resolved. Herschel thus put a question mark against the nature of the object – the first person to do so. When he observed it further, Herschel saw that the object had a real motion with respect to the Earth. He therefore concluded that he was looking at a comet!

Several months were spent trying to fit the new 'comet' to a suitable cometary orbit, until Lexell suggested that the orbit was probably planetary. Once the suggestion had been made, it was at once seen to be obvious. As Kuhn put it, 'A celestial body that had been observed off and on for almost a century was seen differently after 1781 because, like an anomalous playing card, it could no longer be fitted to the perceptual categories (star or comet) provided by the paradigm that had previously prevailed.'

Kuhn points out that the discovery of Uranus did more for astronomy than merely add another planet to the solar system. It prepared astronomers to perceive other such objects, and after 1801 they did indeed begin to see numerous minor planets and asteroids. No fewer than twenty such planetary bodies were discovered by astronomers using standard instrumentation in the first fifty years of the nineteenth century.

This failure simply to see what is before our eyes is far from rare. In the 1890s, scientists all over Europe were experimenting with cathode rays – electrons accelerated in a partially evacuated tube by an electric charge. Researchers trying to tease out the secrets of cathode rays included great names such as Lord Kelvin, who had contributed to the mathematical foundations of electricity and magnetism including the electromagnetic theory of light.

One of these hopeful experimenters was the young Wilhelm Roentgen, working at the University of Wurtzburg in 1895. One day, Roentgen noticed that a screen near his shielded cathode-ray apparatus glowed when the cathode rays discharged. Roentgen locked himself in his laboratory virtually night and day for many days before emerging to announce the discovery of X-rays. By the time he unlocked his laboratory door he had discovered that the new rays travelled in straight lines, that they cast shadows and could not be deflected by a magnet.

When Roentgen announced his discovery it was greeted with surprise and with shock. Lord Kelvin pronounced X-rays an elaborate hoax.[7] Other scientists, though they felt bound to accept the physical results, were staggered by the discovery. Yet, as Kuhn points out, the discovery of X-rays was, 'not at least for a decade after the event, implicated in any obvious upheaval in scientific theory':

> To be sure, the paradigm subscribed to by Roentgen and his con-
> temporaries could not have been used to predict X-rays. (Maxwell's
> electromagnetic theory had not yet been accepted everywhere, and
> the particulate theory of cathode rays was only one of several current
> speculations.) But neither did those paradigms, at least in any
> obvious sense, prohibit the existence of X-rays. . . . On the contrary,
> in 1895, accepted scientific theory and practice admitted a number
> of forms of radiation – visible, infrared and ultraviolet. Why could
> not X-rays have been accepted as just one more form of a well-
> known class of natural phenomena? Why were they not, for
> example, received in the same way as the discovery of an additional
> chemical element? New elements to fill empty places in the periodic
> table were still being sought and found in Roentgen's day. Their
> pursuit was a standard project for normal science, and success was
> an occasion only for congratulations, not for surprise.[8]

There can be little doubt that many European scientific laboratories must have been producing X-rays on a substantial scale yet no one had perceived them. Anyone who thinks that this is merely a case of people 'not noticing' the new rays should remind themselves that Britain's most eminent physical scientist, Lord Kelvin, declared them to be a hoax. There is more to this than not noticing.

Interestingly, at least one other eminent scientist was on the track of X-rays: Sir William Crookes, who had been alerted by some photographic plates that had become unaccountably fogged while covered up. Crookes's exceptional openness to the possibility of a new form of radiation may perhaps be connected with his high tolerance to dissonant ideas – a trait which he demonstrated repeatedly in his later researches.

In many of the examples given so far, we are looking back in time and examining cases of fundamental scientific importance. But how does this strange phenomenon affect ordinary working scientists today? The answer is that it affects them in exactly the same way that it affected Roentgen and Lord Kelvin. Scientists at Oak Ridge, Los Alamos, Stanford University, US Naval Laboratory and Texas A & M University have built Fleischmann-Pons cold fusion cells and they have perceived gamma-rays, tritium and excess heat energy. Scientists at Harwell and MIT have built Fleischmann-Pons cells and have gone on record as saying that they do not see such results – one eminent MIT scientist even claiming, in the best traditions of Kelvin, that cold fusion is a hoax.

Researchers at Stanford Research Institute, Birkbeck College and King's College say they have perceived (and, indeed, filmed and recorded) people producing readings on electrical instruments remotely without touching them, by means that are inexplicable. Researchers at other institutions say they have been unable to perceive or record such things and that the results must be conjuring tricks.

Most extraordinary of all, we have cases where the *same* scientist says that he perceives paranormal phenomena on one occasion, but is unable to see the same phenomena produced by the same individual on a later occasion – as in the case of Dr John Taylor and Uri Geller. No one could accuse Dr Taylor of not being open-minded on the subject. Quite the contrary, he has risked his reputation with a courage and pioneering spirit that has left most of his colleagues gasping. What strange force is it then, that can cause even the most fearless and objective of researchers to undergo such dramatic changes in perception?

The failure to 'see' experimental results sometimes comes about because our expectations direct our attention to the wrong place. Otto Hahn and his colleague Fritz Strassman are famous for their experimental work that led to the discovery that uranium atoms could split apart, turning into other elements – the basis of all nuclear fission discoveries. But after five years' hard work in the 1930s looking for experimental evidence of this process they almost missed it entirely because they were looking for the wrong fission products. As uranium is a very heavy element, they expected the uranium atom to break into other heavy elements, such as radium, thorium and actinium. Actually what they should have been looking for chemically were light elements from the other end of the periodic table: barium and krypton.

The gas krypton was not identified by chemical means until the fission reaction was already well understood, and the second main fission product, barium, was discovered merely by chance because the researchers were adding barium to their radioactive solutions to try to precipitate the heavy elements they were looking for. When they found more barium than they were putting in themselves, they realised something strange was happening.

Hahn himself appears to have suspected that some unknown influence was at work when he wrote:

> As chemists we should be led by this research . . . to change all the names in the preceding [chemical reactions] and thus write barium, lanthanum and cerium, instead of radium, actinium, thorium. But as 'nuclear chemists' with close affiliations to physics, we cannot bring ourselves to this leap which would contradict all previous experience of nuclear physics. It may be that a strange series of accidents renders our results deceptive.[9]

It is interesting to compare Otto Hahn's comments above ('this leap which would contradict all previous experience of nuclear physics') with those of Paul Henri Rebut, director of fusion research at Culham, commenting on Fleischmann and Pons's discovery of cold fusion: 'To accept their claims one would have to unlearn all the physics we have learnt in the last century.' Hahn decided to risk the 'contradiction' and, as a result, discovered nuclear fission.

When we go back again to the psychology laboratory seeking further enlightenment on the nature of the 'strange series of accidents' of which Hahn wrote, we find further experiments suggesting that it is not merely our perception of the contents of a test-tube that can change, but our whole world view. It was as long ago as 1897 that George Stratton first performed an experiment that has become familiar today. An individual who is fitted with a pair of goggles containing inverting lenses is at first completely disoriented by the unaccustomed view of an upside-down world. But after the subject has learned to deal with his new view of the world his entire visual field adjusts itself to the inverted input. After a period of confusion, the subject sees the world 'right way up' again.[10] 'Literally as well as metaphorically,' observes Thomas Kuhn, 'the man accustomed to inverting lenses has undergone a revolutionary transformation of vision.' In the jargon of experimental psychology, he has experienced a 'Gestalt' switch.[11]

Often it is easier for a scientist from a different field – a different world as it were – to see and understand the implications of experimental results. This was the case with John Dalton, originator of the chemical atomic theory. Surprisingly, Dalton was not a chemist and had no special interest in chemistry at first. He was a meteorologist who wanted to understand weather patterns and who concluded that to do so he must familiarise himself with the way in which gases mix and are absorbed by water. He therefore approached the problem with a paradigm very different from that of his contemporaries who were physical chemists. The ruling paradigm for men like Berthollet and Guy-Lussac, Richter and Proust, was that chemicals had a certain affinity for one another.

To Dalton, the pragmatic weather investigator, the mixing of gases and liquids were simply physical processes in which affinity played no part. Thus Dalton took the fact – already known – that some chemical compounds contained fixed proportions of substances and merely generalised it to include all compounds.

Dalton's conclusions were widely attacked by chemists, especially Berthollet who never accepted the atomic nature of chemical elements. But the new generation of chemists, not committed to the old paradigm, was more receptive.

According to Thomas Kuhn:

> What chemists took from Dalton was not new experimental laws but a new way of practising chemistry (he himself called it the 'new system of chemical philosophy'), and this proved so rapidly fruitful that only a few of the older chemists in France and Britain were able to resist it. As a result chemists came to live in a world where reactions behaved quite differently from the way they had before.

The human phenomenon we are dealing with – though rather worrying and disturbing to our normal world view – has so far still been dealt with in a rational, reductionist kind of way. Test subjects misperceived anomalous playing cards in much the same sort of way that rats learn to navigate mazes and dogs salivate when they hear their food bell. But is there anything really important in all this? Is changing the colour of playing cards merely an amusing trick, or does it tell us something more important, more fundamental about the way in which we see and understand the world? Kuhn concluded that it does.

Consider the paradigm shift that occurred in the late Middle Ages when Galileo's view of the pendulum replaced that of Aristotle's. To Aristotle and his contemporaries, heavy bodies possessed a tendency to move by their own nature from higher positions to a state of natural rest at a lower position. Thus they considered that a weight on the end of a chain was merely a weight that was being prevented from falling properly, and achieved its state of rest only after tortuous motions attempting to gain the lowest position. Galileo, says Kuhn, looked at the swinging body and saw a pendulum, a body that almost succeeded in repeating the same motion over and over again *ad infinitum*.

'Having seen that much,' he says, 'Galileo observed other properties of the pendulum as well and constructed many of the most significant and original parts of his new dynamics around them.'

It was from the pendulum, for instance, that Galileo got his only really sound argument to support his view that the rate at which bodies fell was independent of their weight (the story of him dropping cannon balls from the top of the leaning tower of Pisa being, sadly, apocryphal). All these things, Galileo saw for the first time, even though such observations had been made for thousands of years.

Importantly, points out Kuhn, the change did not occur because Galileo was able to make more accurate measurements, or because Galileo was more 'objective'. On the contrary, the Aristotelian description of the pendulum is just as accurate.

Galileo's individual genius is, of course, a key factor in the discovery. But it was not a genius for measurement but a genius for perception. And, interestingly, Galileo had not been educated entirely in the traditions of Aristotle, but had also been exposed to a medieval paradigm of which little trace remains today except the word 'impetus'. Fourteenth-century scholars Jean Buridan and Nicole Oresme formulated the theory that the continuing motion of a heavy body is due to an internal power implanted in it (impetus) by the projector that initiated its motion. Oresme wrote an analysis of a swinging stone in what now appears as the first discussion of the pendulum.

Oresme's view, says Kuhn, 'is clearly very close to the one with which Galileo first approached the pendulum. At least in Oresme's case, and almost certainly in Galileo's as well, it was a view made possible by the transition from the original Aristotelian to the scholastic impetus paradigm for motion. Until that scholastic paradigm was invented, there were no pendulums, but only swinging stones, for the scientist to see. Pendulums were brought into existence by something very like a paradigm-induced Gestalt switch.'

So here we have evidence of a real change in world view taking place and caused by a change of paradigm. But Kuhn has been saving up a much more worrying question for us.

> Do we, however, really need to describe what separates Galileo from Aristotle, or Lavoisier from Priestley, as a transformation of vision? Did these men really *see* different things when *looking* at the same sorts of objects? Those questions can no longer be postponed, for there is obviously another and far more usual way to describe all of the historical examples outlined above. Many readers will surely want to say that what changes with a paradigm is only the scientist's interpretation of observations that are themselves fixed once and for all by the nature of the environment and of the perceptual appar-

atus. On this view, Priestley and Lavoisier both saw oxygen, but they interpreted their observations differently; Aristotle and Galileo both saw pendulums, but they differed in their interpretations of what they both had seen.

Let me say at once that this very usual view of what occurs when scientists change their minds about fundamental matters can be neither all wrong nor a mere mistake. Rather it is an essential part of a philosophical paradigm initiated by Descartes and developed at the same time as Newtonian dynamics. That paradigm has served both science and philosophy well. Its exploitation, like dynamics itself, has been fruitful of a fundamental understanding that could perhaps not have been achieved in another way. But as the example of Newtonian dynamics also indicates, even the most striking past success provides no guarantee that crisis can be indefinitely postponed. Today research in all parts of philosophy, psychology, linguistics, and even art history, all converge to suggest that the traditional paradigm is somehow askew. That failure to fit is also made increasingly apparent by the historical study of science to which most of our attention is necessarily directed here.

The important point here is that what happens during a scientific revolution cannot be reduced simply to a reinterpretation of individual data that remain stable before and after that revolution. As Kuhn points out, a pendulum really is not a falling stone. Oxygen really is not 'de-phlogisticated air' (as some before Lavoisier thought). So the data that scientists collect from their observations actually are different.

'More important,' he concludes, 'the process by which either the individual or the community makes the transition from constrained fall to the pendulum, or from de-phlogisticated air to oxygen is not one that resembles interpretation. How could it do so in the absence of fixed data for the scientist to interpret? Rather than being an interpreter, the scientist who embraces a new paradigm is like a man wearing inverting lenses. Confronting the same constellation of objects as before and knowing that he does so, he nevertheless finds them transformed through and through in many of their details.'

The research work reviewed briefly here seems to me to point to a single unequivocal conclusion: that the human mind plays an active role in the process of perception. The mind is no mere passive mirror reflecting external events. It does not merely represent data in the way that a computer monitor does on a 'dot for dot' basis. Instead it contributes something to the sensory information presented to it. The something that it contributes comes from our existing experience, and the nature and meaning of our existing experience includes the consensus view that we strive to reach to reduce cognitive dissonance to a minimum.

Put at its simplest, what we perceive when we make our observations depends at least in part on what we already believe is there. This in itself carries the disturbing implication that any form of scientific research may be susceptible to an exceptionally subtle form of systematic bias. But there are strong indications from other research that there may be even more powerful forces at work distorting our observations.

CHAPTER NINE

The Research Game

*The heart has its reasons
which reason knows nothing of.*
BLAISE PASCAL
Pensées

Among the many examples of the rejection of valuable inventions and discoveries examined earlier, a sizeable number come from the armed forces. When officials at the British War Office rejected de Mole's proposals for a tracked armoured vehicle in 1912 and 1915 they were following the same great tradition that led their Lordships of the Admiralty to reject Parsons's turbine engine and Pollen's automated range-finding system.

Psychologist Norman Dixon, of University College London, cites such rejections as among hundreds of examples of what he graphically called 'the psychology of military incompetence' in his 1976 book of that title.[1]

Dixon's thesis, for which he presents an overwhelmingly strong case, is that there is a systematic form of military incompetence that arises not simply because of the tendency to 'muddle through' or because of the usual accidents, errors and omissions that plague any military undertaking. It arises because military organisations 'make for military incompetence in two ways – directly, by forcing their members to act in a fashion that is not always conducive to military success, and indirectly, by attracting, selecting and promoting a minority of people with particular defects of intellect and personality'.

Dixon points out that it makes sense to speak of military incompetence in a generalised way because historical example shows that where military commanders are conspicuously incompetent, they frequently fail in the same characteristic way and exhibit the same 'symptoms' of incompetence. In support of this view Dixon lists fourteen key areas that military incompetence involves:

1 *A serious wastage of human resources and failure to observe one of the first principles of war – economy of force.* This failure derives in part from an inability to make war swiftly. It also derives from certain attitudes of mind . . .

2 *A fundamental conservatism and clinging to outworn tradition,* an inability to profit from past experience (owing in part to a refusal to admit past mistakes). It also involves a failure to use or tendency to misuse available technology.

3 *A tendency to reject or ignore information* which is unpalatable or which conflicts with preconceptions.

4 *A tendency to underestimate the enemy* and overestimate the capabilities of one's own side.

5 *Indecisiveness* and a tendency to abdicate from the role of decision maker.

6 *An obstinate persistence in a given task* despite strong contrary evidence.

7 *A failure to exploit a situation* gained and a tendency to 'pull punches' rather than push home an attack.

8 *A failure to make adequate reconnaissance.*

9 *A predilection for frontal assaults,* often against the enemy's strongest point.

10 *A belief in brute force* rather than the clever ruse.

11 *A failure to make use of surprise* or deception.

12 *An undue readiness to find scapegoats* for military setbacks.

13 *A suppression or distortion of news* from the front, usually rationalised as necessary for morale or security.

14 *A belief in mystical forces* – fate, bad luck, etc.

I believe that the examples presented in the first part of this book demonstrate that the systematic incompetence identified by Dixon is not restricted to military hierarchies but also applies to other hierarchical organisations that selectively attract and promote a certain type of personality. In particular, I believe there is evidence to show that it occurs in institutional scientific research.

In this chapter I wish to present evidence for this assertion. To do so, let me first follow Dixon once more in examining the kind of personality who is most likely to be attracted to a hierarchical organisation – the authoritarian personality.

A great deal is known about the authoritarian personality. Much of this knowledge derives from work conducted in England and America in the years following the end of the Second World War in an attempt to gain some insight into the most extreme form of authoritarianism – that of those who had joined the Nazi party and followed Hitler so blindly.

A study by psychologists at Berkeley University in California found two contrasting personality types.[2] In their most extreme polarised form these two types were represented by the authoritarian personality and the democratic personality. The authoritarian type, according to the study, is characterised by being rigid, intolerant of ambiguity, and hostile to people or groups racially different from himself or herself. Their polar opposite is individualistic, tolerant, democratic, unprejudiced, egalitarian. The researchers arrived at this distinction by interviewing more than 2,000 Americans from all walks of life. Similar results have been obtained from studies in England by Hans Eysenck.[3]

The aim of this post-war American study was to try to determine what made people incline towards fascism and it tested people's attitudes to measure anti-Semitism, ethnocentrism, political and economic conservatism, and anti-democratic or fascist tendencies. It was thus called by them, the 'F' scale.

In short, what the F scale measures is an individual's predisposition towards:

- *Conventionalism* – rigid adherence to conventional middle-class values.
- *Authoritarian submission* – a submissive, uncritical attitude towards the idealised moral authorities of the group with which he or she identifies.
- *Authoritarian aggression* – a tendency to be on the look-out for and to condemn, reject and punish people who violate conventional values.
- *Anti-intraception* – opposition to the subjective, the imaginative and the tender-minded.
- *Superstition and stereotypy* – a belief in magical determinants of an individual's fate, and the disposition to think in rigid categories.

- *Power and 'toughness'* – a preoccupation with the dominance–submission, strong–weak, leader–follower dimension, identification with power figures, over-emphasis upon the conventionalised attributes of the ego, exaggerated assertion of strength and toughness.
- *Destructiveness and cynicism* – generalised hostility, vilification of the human.
- *Projectivity* – the belief that wild and dangerous things go on in the world; the projection outwards of unconscious emotional impulses.
- *Puritanical prurience* – an exaggerated concern with sexual 'goings-on'.

Those individuals who scored very high marks on the degree of their intolerance of other racial groups were interviewed in depth by the researchers to try to find out why they were so prejudiced.

According to Norman Dixon:

> The results delineated the authoritarian personality. People who were anti-Semitic were also generally enthnocentrically prejudiced and conservative. They also tended to be aggressive, superstitious, punitive, tough-minded and pre-occupied with dominance-submission in their personal relationships. That this cluster of traits suggested a unique underlying personality-structure was borne out by the clinical interviews. It seems that authoritarians are the product of parents with anxiety about their status in society. From earliest infancy the children of such people are pressed to seek the status after which their parents hanker.[4]

As Norman Dixon observes, to those not previously versed in the psychology of authoritarianism, the list of traits described earlier may be something of a surprise. Surely orderliness, tough-mindedness, obedience to authority and the rest are the very qualities that we want in our fighting men? Unfortunately, says Dixon, 'as represented in the authoritarian personality they are only skin deep – a brittle crust of defences against feelings of weakness and inadequacy. The authoritarian keeps up his spirits by whistling in the dark. He is the frightened child who wears the armour of a giant. His mind is a door locked and bolted against that which he fears most: himself.'

As one of hundreds of examples of the authoritarian personality in action, Dixon cites a conversation quoted by Admiral Dewar.[5] The conversation takes place in the captain's cabin aboard ship;

> *Lieutenant:* I have prepared a report, sir, on our new fire control organisation with sketches of the voice-pipe arrangements. It may be useful to other ships, and I thought you might like to submit it to the Commander-in-Chief.
> *Captain:* That's good. Do you know who is the controller?
> *Lieutenant:* Yes, sir. Captain Jackson.
> *Captain:* Do you know that he was president of the committee that sat on the approval of the existing voice-pipe communications in HM ships?
> *Lieutenant:* No, sir, but I suppose he will be interested in reading the report.
> *Captain:* I am afraid that I cannot forward a report which suggests that the arrangements which he approved are unsatisfactory.
> *Lieutenant:* The report shows how they can be improved.
> *Captain:* Yes, but I am not going to tell him so.

Compare this military conversation which took place at the time of the First World War with one which took place between the Professor of Psychology at a British University and one of his post-graduate students in 1988:

> *Student:* There seems to be a need for good quality research in the field of hypnosis.
> *Professor:* I would never allow it in my department.
> *Student:* Why not?
> *Professor:* Because hypnosis is not a respectable field for research.
> *Student:* Why not?
> *Professor:* Because it has no serious published literature.
> *Student:* Why is there no literature?
> *Professor:* Because nobody has done the research.
> *Student:* Why has nobody done the research?
> *Professor:* Because it's not a respectable field of research.

The conversation was reported by Dr Ashley Conway, a Harley Street psychologist who believes that hypnosis has a role to play in medical therapy. It was first published in an article written by Conway in the professional journal *Complementary Medicine* and entitled 'The Research

123

Game: A view from the field', an article that caused something of a furore when it was published.[6]

Conway's starting-point is the work of Dixon and that of Eric Berne on games theory as applied to adult social interactions. Significantly, however, Conway applies the theory not to the behaviour of military personnel but to that of scientists – and specifically scientists engaged in medical research.

The idea of examining and understanding complex human interactions in terms of games was made popular by Eric Berne. Conway says, 'The important ingredient of the game metaphor is that there are at least two levels of operation – the explicit/overt and the ulterior/covert. Analysing real life events in terms of a game is not intended to trivialise them in any way. As Berne points out, psychological games can result in suicide, murder, war.'

Why, asks Conway, should the serious business of scientific medical research be comparable to a game? The answer, he says, lies in two interdependent factors: the nature of the work as it is currently practised, and the nature of those who are attracted to many types of research – the two same ingredients that made Norman Dixon wonder about military organisations.

According to Conway:

> The day-to-day work is essentially repetitive, with much attention paid to fine detail. Experiments are structured similarly, whether they examine anaesthetised cats or wakeful humans. They minimise the differences between organisms, and it is assumed that one group (the 'experimental' group is comparable to another (the 'control' group). People become 'subjects', and their feelings about being part of the study are not normally considered relevant. Tests are carried out in as objective and scientific a way as possible, with subjects receiving standardised experimental manipulations . . . Research is therefore objective, depersonalising and largely dehumanising. Results must be clean of contaminants and statistically sound. Most of this research is carried out in universities or teaching hospitals, which are hierarchical in their structure. The type of person attracted to this work environment is most likely one who enjoys the security of a set social and intellectual structure, and has an overriding concern for attention to detail.

Conway also cites an article by Dr David Horrobin in which he identified two different kinds of expert. The first is the expert who knows how to do something useful and important – build a bridge or carry out surgery. The second is an expert on the causes of cancer, or unemployment. This Type 2 expert is not a person who knows how to solve problems, but is an 'expert' by virtue of knowing a lot.[7]

Conway proposes that medical research has today become dominated by the second type of expert, and that research organisations have developed effective ways of responding to the special needs of their members, to ensure that the structure becomes self-maintaining. The more a member gets out of it, the greater his or her vested interest in its preservation.

Should we be concerned about this development? Horrobin thinks we should. He points out that in the last twenty years there have been no substantial reductions in morbidity or mortality associated with major diseases that can be attributed to public funding of medical research.

How can it be possible, asks Conway, that a system that spends millions of pounds a year should apparently produce so little? His answer is that very few of the talented young minds who might produce valuable results will ever get a research grant unless they can demonstrate expertise as a Type 2 expert: a textbook expert. Becoming this kind of expert will be the very factor that makes them less likely to do research that is really useful.

What exactly is it that drives people to make research into a game? Says Conway;

> There are a number of incentives to playing the Research Game: a structured hierarchical system, providing status externally (via a group identity with intellectualism and elitism); room for status growth internally (via appreciation of the peer group), and, perhaps more important, security. The security provided is practical, because it provides a safe career, and the Game could go on for ever, and also emotional; you know where you are with the Game – it reduces uncertainty, maintains equilibrium and the chosen frame of reference, blocks intimacy and keeps an emotional distance which achieves the important effect of 'making' people predictable.

The Game enables the Type 2 expert to avoid taking risks, and therefore to avoid being wrong. It enables the environment to be shrunk to fit a more comfortable perception, rather than demand that the researcher should have to open his mind to all the complexities of human nature.

Conway even goes so far as to list the rules of the Game. For the benefit of those who might want to join in, either as participants or as spectators, here they are:

1 *The player's first obligation is to maintain the status quo of the Game. Within the Game, the object is to rise above the other players.*

2 *Do not research an entirely new field.* [As an example Conway quotes the conversation with the Professor of Psychology quoted earlier.]

3 *It does not matter if the ground explored is old or obvious. The way the Game is played is much more important than the practical significance of the research results.* To give an example, Maier and Seligman[8] review the literature on research into the phenomenon of learned helplessness. The article is 44 pages long and contains 6 pages of references to research which often involves the infliction of considerable suffering on laboratory animals. How do they define their theory? 'When an organism is faced with an outcome that is independent of his responses, he sometimes learns that the outcome is independent of his responses. This is the cornerstone of our view and *probably seems obvious to all but the most sophisticated learning theorist.*' It is fine for the Type 2 expert to throw common sense out of the window, and discover the obvious; it reduces his chances of making a mistake. (Italics added.)

4 *Doing something that is useful outside the Game is not only irrelevant, but seriously discouraged.* One player, finding a solution for a real-life problem seriously disrupts the Game. Other players' 'expertise' is lost, and such a result may give rise to embarrassing questions about why other researchers are not doing something useful too.

5 *Objectivity must be maintained at all times.* Players should not be too concerned for the health or well-being of their subjects. This could become a contaminant to objective observation, but more importantly may bring emotion into the Game, one of the very problems that it has developed to avoid.

6 *Players must at all times show detached coolness for their research.* Enthusiasm for a research goal is discouraged; it promotes the

expectation of useful results, and again it introduces feeling into the Game.

7 *Research should always be seen to be good science.* If only it were as straightforward as Newtonian physics! Still, one of the advantages of the Game is that it enables those playing to pretend that it is. In this way formal scientific hypotheses may be tested, and then subjected to a full and statistical analysis, which is highly regarded because it is so clean and neat and reliable, compared to the rather messy business of trying to sort out people.

8 *Players must at all times use correct language.* The creation of a Game language is an important way of maintaining the feeling of being elite. The language rules, together with complicated statistical analysis of data, ensure that the illusion of real expertise is maintained.

9 *The rules of all psychological games are unspoken. Never make the rules of the Game explicit, even to oneself.*

Of course, there are many competent medical researchers producing useful results. Conway says his concern is that the ratio of good, useful, innovative research to 'expertly' designed research of little practical use is out of proportion.

He describes one mechanism by means of which the findings of any really useful research can be ignored or blocked; the defence of finding fault with minor details as a way of discounting the importance of something new.

American psychologist Erik Peper travelled to Europe to attend an international conference on respiratory physiology. His contribution to the conference was to describe his new methods of teaching asthma sufferers how to adjust their breathing in order to stop asthma attacks. Conway describes what happened next.

> Was he greeted with open arms? Could we tell asthma patients that soon they may not have to be dependent on potentially harmful drugs? No, Peper was a threat. Not only was he a Type 1 expert (he was getting people better) but he had broken a number of important rules. He had demonstrated a solution to the problem of how to treat asthma, and this could put a number of 'asthma experts' (let alone drug companies) out of business. He talked to patients and interacted with them as people; he treated people as individuals. All

of this threatens the Game. It is a worrying demonstration of what can happen if enthusiasm and practical ability are combined. He therefore needed to be dealt with quickly. An interested lay person might have asked questions like 'Does this work for everybody? Is it of some help even if it does not offer a complete removal of this condition? Can you teach me how to do it?' Instead Peper was asked detailed questions about the type of electrodes he used in electro-myograph (EMG) biofeedback, which was a small part of his treatment procedure for some of his patients. Eventually, by examining his EMG technique microscopically, the assembled Type 2 experts found something to fault and were able, on the basis of this finding, to reassure themselves that they could go back to the Game and ignore his data.

Conway points out that those graduates who enter research and encounter the rules of the Game will have to have a very strong constitution to stay in research and not become drawn in by the self-reinforcing system. Not surprisingly, most graduates who want to make a real contribution to research and are not happy with the Game are likely to drop out.

So far, we have considered only the consequences of the research Game in human terms. But, of course, as far as medical research and many other forms of scientific research are concerned, there is also the question of money. Conway points out that the ability to play the Game is ultimately determined by hard cash. The direction of gov-ernment funding is determined by those who know most about such research – those who have risen to the top of the Game and hence are presumably the most skilful players. Even independent charities have to bring in 'expert' advice on where to allocate their funds. No prizes for guessing what sort of experts they often call on.

Horrobin asks why we find 'expert committees repeatedly and consistently refusing to support highly innovative lines of research. The Type 2 expert has a vested interest in the answer to the problem not being found. All experts in research on unsolved problems will lose prestige, grants and even incomes when those problems are solved . . . if the research cannot be done, then the solution cannot be demon-strated and the [Type 2] expert can remain safe in his faulty expertise.'

Is there anything that we can do to remedy this state of affairs? Ashley Conway thinks that what is needed is radical change.

> I believe this change should be in both funding and in the choice of who is doing what. *Every* piece of grant-sponsored research should be demonstrably aimed at doing something useful for patients, and not merely another piece of intellectual exhibitionism. At present we have a system where medical and psychological research is frequently carried out by people who do nothing else but research, and are not involved in getting people well. Radical research should be carried out involving experts who work with people at the sharp end. The primary goal of researchers should not be the enhancement of their own status as individuals and as a group, but a real advance in useful, practical knowledge.

Looking at scientists under the microscope of psychological research is informative in one sense but, equally, frustrating in another. Close observation tells us much about some of the hidden or unacknowledged factors that may predispose some scientists to the taboo reaction in their work; but it fails to tell us anything about why some of our fellow men and women may give way to such anti-scientific impulses, when they of all people might be expected to be immune to them.

Is it possible to gain some additional insight by referring to the mountains of anecdotal material that exist in the form of biography, autobiography and even fiction? In opening up this line of investigation I am acutely aware that I am stepping outside the relative safety of reporting the findings of professional studies and strolling into a minefield of conjecture and speculation. I think the risk worthwhile, because there is much valuable literature from the non-research world that offers considerable insight into the taboo reaction.

In the field of fiction, C.S. Lewis has pointed out that not only are scientists vulnerable to the vanities to which we are all prey but that their chosen profession offers temptations that are sometimes exceptionally hard to resist. Institutional science, engaged on momentous projects of national importance, can stimulate a natural longing for a special sense of power: not the sense of power that comes from acknowledging the salute of goose-stepping ranks of soldiers or even the applause of the party faithful at the annual political conference. This is a secret power: the power that proverbially comes from the possession of knowledge; the secret power of being on the inside, being a party to great truths and great deeds, part of important national institutions, while the rest of humanity merely puts on the kettle and makes a cup of tea.

Lewis captured this natural longing perfectly in the character of Mark Studdock, the hero of his science fiction novel, *That Hideous Strength*. Studdock, a young and healthily ambitious sociology fellow at a minor university has been offered a job with the prestigious National Institute of Co-ordinated Experiments and is being driven there to meet the director by no less than Lord Feverstone, something of a demon motorist.

> The long straight nose and the clenched teeth, the hard bony outlines beneath the face, the very way he wore his clothes, all spoke of a big man driving a big car to somewhere where there would be big stuff going on. And he, Mark, was to be in it all.[9]

Once Mark gets to work on the 'big stuff' at NICE he quickly learns to conform.

> Statistics about agricultural labourers were the substance: any real ditcher, ploughman or farmer's boy was the shadow. Though he had never noticed it himself, he had a great reluctance, in his work, ever to use such words as 'man' or 'woman'. He preferred to write about 'vocational groups', 'elements', 'classes', and 'populations': for, in his own way, he believed as firmly as any mystic in the superior reality of the things that are not seen.

These forces, the desire to be part of big stuff, a belief in the superiority of scientific ideas over the ugliness and pettiness of real people, gradually corrupt Mark until without even noticing it, he comes to subscribe to the beliefs and tenets of his inner-circle group even when they become cruel, inhuman and ultimately criminal. People outside the group are ridiculed and treated with contempt while those insiders who show signs of rebelling are degraded and tried on trumped-up charges; but it is in the name of, and for the ultimate benefit of mankind: the greatest good of the greatest number.

This may be merely science fiction, but contrast the fictional Mark Studdock with the very real scientists who worked on the Manhattan Project to develop the first atomic bomb. When the time came in 1945 to test the first bomb in the desert of New Mexico, distinguished physicist Philip Morrison personally collected the components of the plutonium core from their high security vault and sat on the back seat of a car with the core beside him as the car drove to the test site at Alamagordo.

'I remember,' he said later, 'when we were driving through Santa Fe, which was then quite a sleepy little town. I was just thinking about what an extraordinary thing it was to be driving along there in just an ordinary car and yet we were carrying the core of the first atomic bomb.'[10] One can almost picture Lord Feverstone, teeth clenched, at the wheel.

After the core had been assembled at a deserted farmhouse, it was driven by another physicist on the project, George Kristiakovsky. 'For reasons of security we transported it at night,' he recalled, 'but to be whimsical I decided that I would start the trip at ten minutes after midnight, Friday the thirteenth.'[11]

Lewis's fictional portrait of life inside institutional science might also be contrasted with that of real research painted by James Watson, Nobel laureate and co-discoverer of the structure of DNA. Writing with exceptional candour, Watson described how he viewed the 'race' to elicit the structure of DNA with eminent American chemist Linus Pauling who was also on the trail. At one stage, Watson goes to a dance with a lady friend and describes how 'The dance floor was half vacant, and even after several long drinks I did not enjoy dancing badly in open view. More to the point was that Linus Pauling was coming to London in May for a meeting of the Royal Society on the structure of proteins. One could never be sure where he would strike next. Particularly chilling was the prospect that he would ask to visit King's.'[12]

A little later Watson writes that, 'Fortunately, Linus did not look like an immediate threat on the DNA front. Peter Pauling [Linus's son] arrived with the inside news that his father was preoccupied with ... the hair protein, keratin.'

A few months later, however, Watson learned from Peter Pauling that his father had written a manuscript intended for publication on the subject of DNA.

> Peter's face betrayed something important as he entered the door, and my stomach sank in apprehension at learning that all was lost. Seeing that neither [Francis Crick] nor I could bear any further suspense he quickly told us that the model was a three-chain helix. . . .
>
> Giving Francis no chance to ask for the manuscript, I pulled it out of Peter's outside coat pocket and began reading. By spending less than a minute with the summary and the introduction, I was soon at the figures showing the locations of the essential atoms.

To his evident relief, Watson discovered that Pauling's DNA model must be flawed because it was not actually an acid at all (DNA is essentially a moderately strong acid).

> When Francis was amazed equally by Pauling's unorthodox chemistry, I began to breathe slower. By then I knew that we were still in the game.
>
> The blooper was too unbelievable to keep secret for more than a few minutes. I dashed over to Roy Markham's lab to spurt out the news and to receive further reassurance that Linus' chemistry was screwy.
>
> Then, as the stimulation of the last several hours had made further work that day impossible, Francis and I went off to the Eagle. The moment its doors opened for the evening we were there to drink a toast to the Pauling failure. Instead of sherry, I let Francis buy me a whiskey. Though the odds still appeared against us, Linus had not yet won his Nobel.[13]

Of course, no one would try to blame Watson for being as human as the rest of us – on the contrary his honesty and his insight are admirable. Watson is regarded as something of a maverick, especially by his more staid scientific contemporaries. He is thought of by many of them as having been imprudent in letting the profane world see what is really under the scientific kilt. But he is far from alone in confessing to the ignoble but all too human impulse of being ambitious for scientific glory.

It is not so much irreverent to assert that serious-minded professional scientists could ever be subject to such feelings, as foolish to attempt to deny such an obvious truth.

Looking back over the evidence from experimental psychology together with the many concrete examples from the history of science, it seems to me that there are a number of forces at work in scientific research that are clearly distinguishable.

There is the human tendency to see what we expect to see; what fits with our beliefs and our expectations. There is a natural tendency to strive to reduce cognitive dissonance – the gulf between what we already believe and the new information we come into contact with that contradicts our existing beliefs. There are scientific paradigms into which new data must fit or risk being rejected. There is the authoritarian personality, intolerant of dissonance and attracted to hierarchical organisations such as those that scientific research often produces. And there are powerful forces motivating such people to act out a game whose objects and methods are inimical to the new and categorically opposed to the paradigm shift. For some individuals there is a deep need to be a part of an important institution or project, a need which in some circumstances can even take precedence over normal humanitarian feelings. And the more elite, the more desirable the membership of such a 'club', the more likely are such individuals to devalue and derogate those outside the group who disagree with their aims or methods. In these circumstances, the question we should perhaps be asking is how do scientific institutions *ever* make any new scientific discoveries?

One answer is that many don't – they only ever work up to and around the edges of the current paradigm and then stop to defend the borders of that paradigm. Those borders appear quite logically to them as being the limits of reason – and what lays on the other side of those borders as being the irrational. Thus do some scientists become the guardians of the 'gates of unreason'.

CHAPTER TEN

Guardians of the Gate

Men do not get what they deserve
but what they resemble.
JACQUES RIVIÈRE

In Chapter 5, I pointed to a pattern that recurs in the investigation of phenomena that are dissonant with the current scientific paradigm, involving the ridicule, ostracism and punishment of the discoverer. A key figure in these proceedings is the actor who casts himself or herself in the role of the people's protector or saviour from the machinations of charlatans, false prophets and weirdos of every sort.

In the seventeenth century this figure would very likely have been a Witchfinder General; today, mercifully, we no longer consign people to the flames to reduce our cognitive dissonance. But, as we saw in Chapter 4, we still burn people's books to achieve the same end. And the present-day book burner turned out to be none other than an agent of the government itself – the US Food and Drug Administration. Many people may draw comfort from the reflection that Wilhelm Reich's books were burned more than thirty years ago in 1960 – it couldn't happen today. Sadly, however, the spirit of Salem is still alive and as recently as 1981 when Professor Rupert Sheldrake published his concept-shattering *A New Science of Life*, the editor of *Nature*, John Maddox, ran an editorial saying the book was 'the best candidate for burning there has been for many years'.[1]

Orwell's Thought Police are all too real. They do not ride sinister black motorcycles nor throw people in jail (although, as we have seen, even that is not unknown). But their powers to punish those who step out of line can be very real. And their effects on the community are no less profound because these individuals are often self-appointed. It is not our political thoughts they police, but the current paradigm.

Today, the Paradigm Police crop up in a variety of unremarkable guises. Most often, it is to style themselves as myth busters of one sort or another or as the guardians of the gates of unreason who we met at the end of the last chapter. Whatever their guise, the Paradigm Police are invariably present wherever innovators and discoverers of the new are derided and attacked. Their methods, their beliefs and their motivations are thus of considerable interest to anyone who wants to understand the taboo reaction in science.

The Paradigm Police are sometimes harmless, because their intemperate behaviour negates any effects they might have. One of the most common characteristics of the authoritarian person is an inability to control or moderate his or her reaction to being confronted by cognitive dissonance. The need to attack the offending agent of dissonance, by any and every means to hand, makes such a person overwhelmingly intemperate and intolerant and gives the game away. But in other cases, the effects of this policing can be very destructive and very far reaching.

One of the best-documented modern examples of the Paradigm Police in action is provided by the case of Dr Immanuel Velikovsky, the American psychologist whose 1950 book *Worlds in Collision* caused a storm of controversy in the US academic world.[2] In 1963, the magazine *American Behavioral Scientist* thought the way in which Velikovsky was mugged by the scientific community of sufficient interest to devote a special issue to three papers on the subject, one by Professor Alfred de Grazia of New York University and two others by Ralph Juergens and Livio Stecchini.[3] The papers, together with additional material were published in book form under the title *The Velikovsky Affair*.[4]

According to de Grazia, Velikovsky's book

> gave rise to a controversy in scientific and intellectual circles about scientific theories and the sociology of science. Dr Velikovsky's historical and cosmological concepts, bolstered by his acknowledged scholarship, constituted a formidable assault on certain established theories of astronomy, geology and historical biology, and on the heroes of those sciences. Newton himself, and Darwin were being challenged, and indeed the general orthodoxy of an ordered universe.

What must be called the scientific establishment rose in arms, not only against the new Velikovsky theories but against the man himself. Efforts were made to block the dissemination of Dr Velikovsky's ideas, and even to punish supporters of his investigations. Universities, scientific societies, publishing houses, the popular press were approached and threatened; social pressures and professional sanctions were invoked to control public opinion. There can be little doubt that in a totalitarian society, not only would Dr Velikovsky's reputation have been at stake, but also his right to pursue his enquiry, and perhaps his personal safety.

As it was, the 'establishment' succeeded in building a wall of unfavourable sentiment around him: to thousands of scholars the name Velikovsky bears the taint of fantasy, science-fiction and publicity.

The central theme of the book that caused such a furore is that between the fifteenth and eighth centuries BC the earth underwent a series of global catastrophes. Parts of the surface were heated until they melted and the seas boiled and evaporated. Some mountain ranges disappeared while others were thrown up elsewhere. Continents were raised, causing global flooding. Velikovsky supported this picture of worldwide catastrophe with a wealth of quotations from such ancient sources as the Hebrew Bible, the Hindu Vedas, Roman and Greek mythology and the myths and legends of many ancient races. He also supported it with physical evidence from geology and palaeontology.

The cause of these tremendous upheavals, according to Velikovsky, was an extraordinary series of astronomical events. He brought forward evidence to suggest that in the past there have been collisions or near collisions between planets in the solar system and that the earth itself experienced a collision with the tail of a comet that ended up as the planet Venus. These events, said Velikovsky, were responsible for repeated changes in the Earth's orbit and the inclination of its axis. Interactions between the magnetic fields of the earth and other planets played a major role in these events.

The story of what happened when the book was published was told by Ralph Juergens in his article 'Minds in Chaos'.[5] Velikovsky first signed a contract for a book on this subject with Macmillan Company in 1946. By 1950, the book was ready for publication. In January of that year *Harper's Magazine* published two articles condensed from the book, under the heading 'The day the Sun stood still', and the magazine was a sell-out. Papers in America and abroad reprinted the articles and further popular articles followed in *Reader's Digest* and *Collier's Magazine*. Most of the articles were highly sensationalised and Velikovsky threatened to disown the articles unless they were toned down.

When these sensational stories caught the public imagination, the scientific establishment began to react. Just before the book was to be published, Macmillan received two letters from Harlow Shapley, professor of astronomy at Harvard University. In the first Shapley described his astonishment that Macmillan should even consider a venture into the 'black arts', but expressed his satisfaction that the publisher had come to its senses and decided not to publish after all. When the firm wrote back to explain that Shapley was the victim of a rumour and that publication was to go ahead as scheduled, Shapley (who had still not seen the manuscript) replied that: 'It will be interesting a year from now to hear from you as to whether or not the reputation of the Macmillan Co. is damaged by the publication of *Worlds in Collision*.' He ended by saying that Velikovsky's background should be investigated as it was quite possible that the book was 'intellectually fraudulent'.

In February 1950, an issue of *Science News Letter* which was edited by Shapley, printed denunciations of Velikovsky's ideas by five scientific authorities in the fields of archaeology, oriental studies, anthropology, geology and with Shapley himself speaking for astronomy. This broadside was published to coincide with the publication of the book – which none of the critics had yet seen.

Perhaps if Velikovsky's book had been of a purely speculative nature then academics would merely have dismissed it as fantasy and not troubled themselves about its content. But Velikovsky backed up his theories with immensely detailed scholarly research in many different disciplines – history, anthropology, geology, astronomy and biology being only some. In fact, he displayed a grasp of his subject that was clearly beyond some of his most vociferous critics, with the predictable consequence that they did not reply to or even address the scientific issues raised, but instead attacked him personally.

In the next few months, newspapers around the country were barraged with abusive reviews contributed by big-name scientists. Virtually none of the reviewers confronted the scientific issues but simply derided Velikovsky. Paul Herget, director of the observatory at the University of Cincinnati, concluded that the book's astronomical ideas were 'dynamically impossible' but offered no reasoned explanation of this conclusion. Californian physicist H.P. Robertson wrote, 'This incredible book . . . this jejune essay [is] too ludicrous to merit serious rebuttal', thus saving himself the trouble of writing any such rebuttal. Nuclear physicist Harrison Brown told the readers of the *Saturday Review of Literature* that the list of errors in fact and conclusion contained in Velikovsky's book would fill a thirty-page letter, although he neglected to specify even one of them.

Despite (perhaps because of) this campaign, the book went to the top of the best-seller list and stayed there for twenty successive weeks. However, in May, when book sales were at their peak, Velikovsky was summoned to Macmillan's offices and told that professors in certain large universities were refusing to see Macmillan's salesmen. This was a serious threat to the company because a substantial part of its revenue derived from the sale of textbooks to universities. In addition, letters had been received from scientists demanding that Macmillan cease publication. Macmillan told Velikovsky that they had no alternative but to respond to this commercial pressure and that they had worked out a deal under which Doubleday would take over publication of the book. Doubleday had few textbook titles and so was relatively immune to academic blackmail.

On 11 June 1950, the *New York Times* carried an article by columnist Leonard Lyons who broke the news.

> The greatest bombshell dropped on Publisher's Row in many a year exploded the other day. . . . Dr Velikovsky himself would not comment on the changeover. But a publishing official admitted, privately, that a flood of protests from educators and others had hit the company hard in its vulnerable underbelly – the textbook division. Following some stormy sessions by the board of directors,

Macmillan reluctantly succumbed, surrendered its rights to the biggest money-maker on its list.[6]

Lyons went on to report that the suppression had been engineered by Harlow Shapley of Harvard, although Shapley later denied this to *Newsweek*. Other scientists were not so shy about admitting their part. Paul Herget said, 'I am one of those who participated in this campaign against Macmillan', while Michigan astronomer Dean McLaughlin wrote, '*Worlds in Collision* has just changed hands . . . I am frank to state that this change was the result of pressure that scientists and scholars brought to bear on the Macmillan Company.'

Even after the change of publisher, ripples of the affair continued to be felt. James Putnam, the editor who had been twenty-five years with Macmillan and who had bought Velikovsky's book, was summarily dismissed. And Macmillan sent a representative to placate the powerful American Association for the Advancement of Science at its annual meeting in Cleveland in December. Charles Skelley, for Macmillan, duly appeared before a committee specially appointed to study means for 'evaluating new theories before publication' − in other words, scientific censorship.

As well as behind-the-scenes pressure on Macmillan, there was also 'nobbling' of senior academics who took Velikovsky's book seriously. According to Alfred de Grazia:

> Several scientists and intellectuals who attempted [Velikovsky's] defence were silenced or sanctioned severely. I. Bernard Cohen, Professor of the History of Science at Harvard University, wrote sympathetically, almost enthusiastically, of Velikovsky's work in the advance summary of his address before the American Philosophical Society in April 1952, but changed his approach markedly in the published version of his address in the *Proceedings of the American Philosophical Society* (October 1952).

Perhaps Professor Cohen was referring to the pressure that had been applied to him to change his mind when he wrote in the same issue of the *Proceedings* his view that, 'Any suggestion that scientists so dearly love truth, that they have not the slightest hesitation in jettisoning their beliefs, is a mean perversion of the facts.'

At the time that Velikovsky wrote, astronomers believed that the planet Venus was an old planet, that its surface was cool like the Earth's, and that its atmosphere consisted largely of water vapour or carbon dioxide. When he had completed the manuscript of the book in 1946, Velikovsky had tried to enlist the help of scientists in conducting experiments that would crucially test his thesis. He made three specific predictions relating to the planet Venus, all of them in principle falsifiable by experiment. First, he said that if Venus were a relatively young planet, its surface temperature would still be very hot. Second, that it would be enveloped in hydrocarbon clouds – the remains of a hydrocarbonaceous comet tail. And third, that it would have anomalous rotation movement, perturbations remaining from its settling relatively recently into orbit.

In 1953, while addressing graduate students at Princeton University, Velikovsky suggested two further testable phenomena: that the Earth's magnetic field reaches as far out into space as the Moon's orbit and is responsible for the libratory or rocking movements of the moon. And he suggested that the planet Jupiter (from which he said the Venus-comet had originated) radiates in the radio frequency range of the electromagnetic spectrum.

These predictions were taken by scientists of the 1950s as being tantamount to proof of Velikovsky's ignorance, insanity or both. Harlow Shapley refused to become involved in any experimental research to confirm his ideas. When, for instance, it was suggested that Shapley might use the Harvard observatory to search for evidence of hydrocarbons in the Venusian atmosphere, Shapley replied that he wasn't interested in Velikovsky's 'sensational claims' because they violate the laws of mechanics and 'if Dr Velikovsky is right, the rest of us are crazy'.

Within little more than a decade of publication, *all* of Velikovsky's key predictions were confirmed by experiment. The *Mariner* spacecraft of 1963 determined that the surface temperature of Venus is in the region of 800 degrees Fahrenheit and that the planet's fifteen-mile thick atmosphere is composed of heavy hydrocarbon molecules and possibly more complex organic compounds as well.

In April 1955, Drs B.F. Burke and K.L. Franklin announced to the American Astronomical Society their accidental discovery of radio noise broadcast by Jupiter. In 1962, the US Naval Research Laboratory in Washington and the Goldstone Tracking Station in southern California announced that radiometric observations showed Venus to have a slow retrograde motion. In the same year, the *Explorer* satellite detected the Earth's magnetic field at a distance of at least twenty-two Earth radii, while in 1965 it was reported that the tail extends 'at least as far as the moon'.[7]

Considering that the main thrust of science's attack on Velikovsky was a personal attack on his integrity, the behaviour of some of his most vociferous critics in the scientific community makes interesting reading. In August 1963, *Harper's Magazine* which had carried the original announcement of Velikovsky's theories, now did a retrospective piece pointing out how all his main predictions had been borne out. The author of both articles, Eric Larrabee, made a reference which drew a thunderous response from Donald Menzel, director of Harvard College Observatory. At the height of the controversy a decade earlier, Menzel had tried to shoot Velikovsky down by calculating that for his astronomical theory to be right, the Sun would have to have a surface potential of 10 billion billion volts. Obviously, said Menzel, this is impossible so Velikovsky must be wrong. By an extraordinary chance, in 1960, V.A. Bailey, emeritus professor of physics at Sydney University (who knew nothing of the Velikovsky controversy) claimed to have discovered that the Sun is electrically charged and has a surface potential of 10 billion billion volts – exactly the value calculated by Menzel.

Feeling that Bailey's discovery made him look foolish, Menzel now sent off a strongly worded response to *Harper's* and a letter to Bailey in Australia asking him to revoke his theory of the electric charge on the Sun as it was assisting the enemy.

According to Ralph Juergens:

> Professor Bailey, taking exception to the idea that his own work should be abandoned to accommodate the anti-Velikovsky forces, prepared an article in rebuttal to Menzel's piece and submitted it to *Harper's* for publication in the same issue with Menzel's. Bailey had discovered a simple arithmetical error in Menzel's calculations, which invalidated his argument.

It is equally interesting to see how the Harvard astronomer dealt with the fact that most of Velikovsky's predictions had been confirmed. On the radio emissions from Jupiter, he wrote that, since most scientists do not accept Velikovsky's theory then it follows that 'any seeming verification of Velikovsky's prediction is pure chance'. As far as the high surface temperature of Venus is concerned, Menzel argued that 'hot is only a relative term'. Later in the article he referred back to this statement saying 'I have already disposed of the question of the temperature of Venus'. Actually, in 1950, Menzel had estimated the temperature of Venus to be about 120 degrees Fahrenheit when it is really more like 800 degrees. On the extent of the Earth's magnetic field, Menzel wrote that Velikovsky 'said it would extend as far as the moon; actually the field suddenly breaks off at a distance of several earth diameters'. In fact, Menzel was wrong; the field had been detected as extending at least twenty-two Earth radii a year earlier by the *Explorer* satellite.

To their credit, a few scientists did support Velikovsky against the climate of hysteria and intimidation including Princeton's Professor H.H. Hess, who was later chairman of the National Academy of Science's space board. In 1962, Princeton physicist Valentin Bargmann and Columbia astronomer Lloyd Motz wrote a joint letter to the editor of *Science* magazine calling attention to Velikovsky's priority in predicting Venus's high surface temperature, Jupiter's radio emissions and the great extent of the Earth's magnetosphere, but *Science*'s editor, Dr Philip Abelson, was not interested in Velikovsky. Instead, he printed a letter from science fiction writer Poul Anderson satirising Velikovsky on the grounds that science fiction writers and hoaxers also made fantastic predictions that were sometimes verified. When the editor of *Horizon* magazine wrote to Abelson protesting at the exclusion of an article by Velikovsky, Abelson replied:

> Velikovsky is a controversial figure. Many of the ideas that he expressed are not accepted by serious students of earth science. Since my prejudices happen to agree with this majority, I strained my sense of fair play to accept the letter by Bargmann and Motz, and thought that the books were nicely balanced with the rejoinder of Anderson.[8]

Scientific American showed that it had not moved on editorially since it ridiculed the Wright Brothers fifty years earlier. The magazine had refused to carry advertising for *Worlds in Collision* and in 1956 it carried a strongly critical article by physicist Harrison Brown. When Velikovsky asked for the right to reply he was told by *Scientific American* editor Dennis Flanagan that:

> I think you should know my position once and for all. I think your books have done incalculable harm to the public understanding of what science is and what scientists do. There is no danger whatever that your arguments will not be heard; on the contrary they have received huge circulation by scientific standards. Thus I feel that we have no further obligation in the matter.[9]

De Grazia highlights an essential issue from this reply when he points out that the editor has picked up a misapprehension common among scientists: that the media of the general public can substitute for the scientific media. Not only is this idea false but, as de Grazia points out, scientists themselves insist upon a distinct separation of the two types of media.

Overall, the attitude of science and scientists during the Velikovsky affair was best summed up by Laurence Lafleur, associate professor of philosophy at Florida State University. Lafleur wrote to *Scientific Monthly* in November 1951 proposing seven diagnostic criteria that would enable anyone to spot the difference between a crank and a scientist. He concluded that Velikovsky qualified as a crank 'perhaps by every one' of them. Lafleur's seven criteria are examined in detail in the next chapter, which is devoted to the question of how to tell a real crank from a real innovator. As far as the present examination of the activities of the Paradigm Police is concerned, the last word should go to Professor Lafleur, since it so accurately sums up the central credo of the 'guardians' of science:

> The odds favour the assumption that anyone proposing a revolutionary doctrine is a crank rather than a scientist.[10]

From anyone concerned with education or science this is a surprising doctrine; from a professor of philosophy it is truly astounding. What Professor Lafleur has defined here is not the nature of *scientific enquiry*; he has defined the nature of *religious heresy*. Rather than evaluation of evidence, we are advised to prefer the assumption that revolutionaries are cranks. Reading Lafleur's dictum through, I am irresistibly reminded of the reason given by Professor John Huizenga for refusing research funding for cold fusion: 'It is seldom, if ever, true that it is advantageous in science to move into a new discipline without a thorough foundation in the basics of that field.' We shouldn't invest in researching things we don't understand, only in things we are thoroughly familiar with.

Under what circumstances can Lafleur ever accept a new discovery? What would have been his reaction to any of the discoveries of the past 500 years – to Copernicus, to Galileo, to Giordano Bruno? Judging by his own yardstick – revolutionary ideas are most likely to be crank ideas – he would have rejected every one of them. Yet he is deluding himself that while in the present he is immune to crank ideas, he would have been receptive and tolerant to these great discoveries of the past.

The story of cold fusion in Chapter 3 introduced us to another self-appointed guardian of the current paradigm in the shape of *Nature* magazine. *Nature* is invariably referred to as the most prestigious, or the most highly respected scientific magazine in the world, and so it is. It is *the* preferred vehicle for announcement of almost every major scientific discovery.

Inevitably, the process of receiving papers from hopeful authors – all of whom would like to see their name in *Nature* and receive the tacit approval of having made it into print in so prestigious a journal – puts the editor and his colleagues into the position of being arbiters of the value of scientific research, whether they wish for such a position or not.

In an attempt to guard against acting in an arbitrary way, and excluding original work of value merely because it covers unfamiliar territory, and at the other extreme, admitting poor or even fraudulent research for publication, *Nature* employs the peer review system. Under this system a number of referees, usually distinguished and reputable scientists in the field in question, are asked by *Nature* to examine the paper, to make comments and to ask for further work or clarification if necessary, and finally to pronounce on its suitability for publication.

Inevitably this will be seen as a kind of 'exam' and rejection will sometimes be perceived by the authors concerned as a 'failure' or perhaps as a form of censorship. In this situation it is equally inevitable that there can be a certain amount of bitterness, allegations of backstabbing, petty jealousies, settling of old scores and other manifestations of the unacceptable face of academia.

The magazine's perception of itself is that it does a reasonably good job at squaring the circle of these conflicting interests given that, like the rest of us, it has limited resources and its staff are only human. The magazine even likes to think of itself as being adventurous when the occasion demands. The question is, what does the record really show?

When *Nature* received the paper written by Drs Harold Puthoff and Russell Targ about their experiments at Stanford Research Institute with Uri Geller in February 1974, they didn't know what to make of it. They were reluctant to get mixed up with Uri Geller and the paranormal, but to reject a paper from two respected physicists at the equally respected Stanford Research Institute would be an insult that would cast doubt on their integrity.[11]

So the magazine adopted a rather different procedure from normal. First it stalled for eight months. However, the paper was being widely circulated in photocopied form and there was much talk about its contents. Scientific gossip hinted that the paper contained categorical proof of the paranormal. *Nature* thus realised that simply to sit on the paper was no solution.

It therefore sent the paper to three referees, as usual. The principal referee it selected for the task, and the man who would later write the editorial that accompanied the paper, was Dr Christopher Evans, an experimental psychologist who worked at the National Physical Laboratory. Dr Evans, who died of cancer in 1979 at the tragically young age of 48, was well known in the 1970s as a writer and broadcaster on popular science topics.

However, his qualifications for being the principal referee of a ground-breaking paper on parapsychology are not immediately obvious. The National Physical Laboratory is a highly respected organisation, but so far as I have been able to discover, it has never conducted any research into the paranormal. Dr Evans was a respected scientist, but again, so far as I can discover he never conducted any research into the paranormal. In fact the only qualification that I have been able to find for the choice of Dr Evans as referee is that a year earlier, in 1973, he had published a book called *Cults of Unreason*, a book that makes interesting reading.[12]

In it, Dr Evans 'exposed' a range of ideas that he said were surrogate beliefs that had sprung up following the decline of orthodox religion. His choice of subjects is revealing. It includes: Scientology, UFOs, 'black box' medical practices, and eastern mysticism. Half the book is about Scientology — then very much in the news and a subject which he had researched in some depth. But the remainder is an odd mixture of stories of flying saucer cults, medical men who claimed to cure with black boxes and the growing influence of eastern esoteric philosophy, such as Taoism, and Zen Buddhism.

Amongst this eclectic bunch there are very probably some people and some groups who deserve to be exposed as charlatans, people interested only in parting others from their money. But the strange thing is that alongside the obvious frauds are people about whom it is at least possible to have a different view if one examines the facts objectively (an activity that was Dr Evans's profession). These people include George de la Warr, Wilhelm Reich, George Gurdjieff, Rudolph Steiner, the Krishna Consciousness movement and others.

There is, of course, no objection to Dr Evans or anyone else exposing these people as fraudulent — so long as they do so by producing some evidence to support such claims of fraud. But that was not Dr Evans's method. Instead, he ridiculed them. He made fun of their beliefs and presented their stories in a way that makes them look slightly deranged. And he included them amongst rather lurid tales of undoubted frauds and con-men so that they appeared guilty by association.

What was it that these people had done to deserve such treatment? They had increased Dr Evans's level of cognitive dissonance by introducing him to startling new ideas — ideas that might contain some important elements of truth. Evans responded in a perfectly natural way. In the words of Leon Festinger quoted earlier, 'Another way of reducing dissonance between one's own opinion and the knowledge that someone else holds a different opinion is to make the other person, in some manner, not comparable to oneself. Such an allegation can take a number of forms. One can attribute different characteristics, experiences or motives to the other person or one can even reject him and derogate him.'[13]

So although, on the face of it, Dr Evans had few qualifications to act as a referee of Puthoff and Targ's paper, he was, from *Nature*'s point of view a perfect choice. And he delivered as expected. The referees came down, on the whole, against publication. *Nature*'s then editor, David Davies, was also against publication – he called it a 'ragbag of a paper' – but he could not reject it outright both because of the reputation of the authors and their research institute and because to reject it would merely stimulate further scientific speculation about the contents of the paper and give credence to allegations of suppression of the SRI results.[14]

Davies then conceived the following plan. He would publish the paper but he arranged with his friend Bernard Dixon, editor of the weekly magazine *New Scientist*, to publish simultaneously an article hostile to Uri Geller. Then he commissioned Christopher Evans to write the editorial comment in *Nature*, which would both be derogatory of the Stanford research and would point readers to the hostile article by Dr Joe Hanlon in *New Scientist*. And this is the plan that Davies and Dixon put into operation.[15]

Evans's *Nature* editorial said that the SRI paper was 'weak in design and presentation, to the extent that details given as to the precise way in which the experiment was carried out were disconcertingly vague'.

> All the referees felt [wrote Evans] that the details given of various safeguards and precautions introduced against the possibility of conscious or unconscious fraud on the part of one or other of the subjects were 'uncomfortably vague' (to use one phrase). This in itself might be sufficient to raise doubt that the experiments have demonstrated the existence of a new channel of communication which does not involve the use of the senses.

In the coded diplomatic language of science, Dr Evans is here telling his readers 'we may have to publish this rubbish, but no-one has to believe it'.

A little later, Dr Evans wrote:

> Publishing in a scientific journal is not a process of receiving a seal of

approval from the establishment; rather it is the serving of notice on the community that there is something worthy of their attention and scrutiny. And this scrutiny is bound to take the form of a desire amongst some to repeat the experiments with even more caution. To this end the *New Scientist* does a service by publishing this week the results of Dr Joe Hanlon's own investigations into a wide range of phenomena surrounding Mr Geller.[16]

Naive readers of these words could be forgiven for imagining that they were being referred to an attempt to replicate, under laboratory conditions, the SRI experiments on Geller, but with tighter controls and greater precautions against fraud. In reality, Dr Hanlon's 'investigation' took the form of two meetings with Geller, one in the lobby of a hotel and the other in the offices of the *Sunday Mirror*. And their scope was restricted to attempts by Dr Hanlon to catch Geller cheating ('I was looking for tricks', he told his readers).

Thus Dr Evans's editorial really translated as: if you want the low-down on how Geller pulls his conjuring tricks, go out and buy *New Scientist* – Joe Hanlon will tell you how he does it. A great many curious people did go out and buy the *New Scientist* that week. But they were disappointed. Hanlon had no scientific evidence at all on Geller, only a mixture of hearsay, gossip, suspicion and speculation that he referred to as 'circumstantial evidence that Uri Geller is simply a good magician.' Since Geller has never been caught cheating in laboratory tests (see Chapter 15), it is a little difficult to see what rational basis there can be for Hanlon's belief – except, perhaps, the deeply held conviction that paranormal phenomena *must* be bogus and, therefore, so must Geller.

This 'assumption of guilt' because the phenomena claimed *must* be fraudulent is a common feature of much writing against those who commit the unpardonable sin of discovering the new, and one which has fuelled many attacks in recent decades. One interesting example on a rather grand scale was provided by a 1981 book entitled *Let's Talk About Me*, jointly written by psychiatrist and writer Anthony Clare, and the producer of his radio series with the same title, Sally Thompson.

The intent of the book can be accurately gauged from this single paragraph at the beginning:

During a period in which psychiatry has become self-obsessed and its public image somewhat blurred and murky, there has been a bewildering proliferation of therapies, alternately described as 'new' and 'fringe'. Such therapies have tended to originate in California and spread east across the United States and thence to Europe. Many of these 'new' therapies stress their orthodox psychiatric roots, others take pride in the extent to which they have thrown off the constraints of established psychiatry and have ploughed genuinely virgin ground. All offer rich prizes – self-discovery, self-perfection, maturity, holism, earthly salvation, a community, a place in the sun.'[17]

In this book, Clare and Thompson have done a number of things and done them well. They have isolated and critically examined each of the main humanistic or development therapy movements that have sprung up over the past thirty of forty years, especially in the United States. They have done their homework carefully on each movement or group and have set those therapies into a common background. They look at such figures as Carl Rogers and the encounter movement, Fritz Perls and Gestalt therapy, Wilhelm Reich and Bioenergetics, Ida Rolf and massage, Jacob Moreno and Psychodrama, Arthur Janov and Primal Therapy, Eric Berne and Transactional Analysis and many more.

They have shown that, in each case, the therapy was initiated by a single powerful personality, that their followers have defended and protected their theories and beliefs in a frankly rather sheep-like way but that in almost every case there is a number of disenchanted customers who have often paid large sums of money, have initially believed that they had received benefits, but ultimately ended up as disbelievers who felt they had been duped.

However, there are also a number of important things that Anthony Clare has not done that are somewhat surprising for a scientist investigating a scientific subject. The first is that he has not set out to interview real users of the therapies except where they happen to fall into his lap – he has used the fly-paper method of collecting views and researching his subjects. Sometimes, the only interview he gives is actually favourable to the therapy.

But whether or not he has a customer interview, and whether or not it is favourable, Clare still arrives at the same conclusion in each and every case that he examines – that the founder was a charlatan, that the movement is a waste of time and money and you, the reader, would be much better off going to a real psychiatrist with real hospital qualifications whose methods can be shown to be scientific – orthodox medical science, in fact.

The second, and more important omission, is that Clare does not appear to have any concrete experience of the therapies he criticises and quite simply has not himself tried them, relying instead on the same sort of critical approach as Christopher Evans.

It is by no means uncommon in taboo fields of research for poachers to turn gamekeepers and vice versa. One well-known case of a scientist changing his mind from being a believer to an unbeliever is provided by Dr John Taylor, whose work was described earlier. In 1974, Dr Taylor examined Uri Geller and other 'sensitives' and gave the following estimate of their significance:

> We can only hope that a careful study of the Geller phenomenon will allow us to reach a better understanding of the wider range of ESP manifestations and any underlying reality that they may expose. In this book we may find that is possible, and arrive at a unified explanation of a large body of ESP phenomena. The question is, can we expect to get from this view of reality any joy for the future of mankind? I think we can.
>
> The way towards this goal will be taken by bringing the entire resources of science to bear on the problem of the nature of the Geller phenomenon. In the process, we may well discover surprising things about the interaction of mind and matter.[18]

By 1980, Dr Taylor had changed his mind and said that: 'Every supernatural phenomenon I investigated crumbled to nothing before my gaze.'[19]

Anyone reading this introduction to his later book might imagine that Dr Taylor must have suffered some pretty dramatic set-backs in his investigations. But consider just one example that he gives in this later book. He first presents details of a well-documented 'poltergeist' case involving a girl of nineteen employed in a law office in the German town of Rosenheim. The girl appeared to be able to cause all manner of strange phenomena over a period of months. Later on he remarks:

> Whatever psychic powers the young girl at Rosenheim may have possessed, she had at least one thing in common with Nina Kulagina, a housewife born in Leningrad in the 1920s, whose ability to make objects move has been investigated extensively by Russian scientists and also by four parapsychologists from the west – H. Keil, B. Herbert, J.G. Pratt, and M. Ullman.
>
> This trait is reported by the western investigators as follows: Kulagina can, by placing her hands on a person's forearm, induce a sensation that feels like very real heat to the point of being painful.... Both Herbert and Fahler had 'burn' marks on their arms which were visible for several hours. No blisters or other negative effects developed.

Of this case, Taylor remarks, 'Direct fraud can be excluded here quite effectively.' The investigators also reported the 'burning' effect when Kulagina did not actually touch the person's arm.

So Dr Taylor has reached his conclusion that paranormal forces are an illusion in spite of evidence such as this that at least two people are independently known who, in controlled conditions that exclude fraud, are able to induce 'burn' marks on the skin of others without direct contact.

Dr Taylor's change of mind is simply puzzling, but the activities of some opponents of the paranormal is alarming. In 1955, Dr George Price of the Department of Medicine at the University of Minnesota published an article in *Science* magazine on the research into extrasensory perception carried out by Dr J.B. Rhine at Duke University and British researcher Dr R.G. Soal. Price argued that ESP was scientifically impossible and that therefore Rhine and Soal *must* be fraudulent experimenters. This authoritative article in so authoritative a journal was interpreted by many scientists as the final nail in the coffin of paranormal research. Nearly twenty years later, in 1972, Price wrote a public apology to Rhine and Soal in the same journal, withdrawing some of his allegations of fraud and admitting that he had made them without even attempting to find any evidence. Price admitted that he had been under the mistaken assumption that Rhine was trying to promote some kind of religious belief.[20]

Some self-appointed guardians of the current paradigm have become so concerned at the modern trend towards 'alternative science' and the distressing tendency for people to think for themselves instead of accepting the received wisdom of science that they have formally created groups dedicated to combating this movement. One such group call themselves CSICOP, the Committee for Scientific Investigation of Claims Of the Paranormal, and have set out to explode what they see as the myth of the paranormal, under the umbrella concept that all phenomena are explicable within the framework of accepted science (although it is not entirely clear how, scientifically, they can know this in advance of actually investigating the phenomena).

CSICOP was founded in the United States in 1976 by Dr Paul Kurtz, a philosopher at the State University of New York, and sociologist Dr Marcello Truzzi, together with a number of other academics. The two principal founders acted as co-chairmen. Scientists in the United Kingdom have also joined CSICOP, amongst the most prominent of which are psychology professor C.E.M. Hansel and Dr Susan Blackmore. The primary aim of the organisation when it was first formed was declared to be, 'the critical investigation of paranormal and fringe-science claims from a responsible, scientific point of view', together with the dissemination of 'factual information about the results of such enquiries to the scientific community and the public'.[21]

This scientific objective, had it been put into practice, would certainly have been welcomed by most paranormal investigators, even if the tone of the investigation was severely critical. However, within a year of formation of the Committee, Truzzi and other members resigned because they were unhappy with the crusading zeal and inquisitional approach adopted by Kurtz and others in place of the spirit of scientific enquiry. Truzzi was replaced with the more ideologically 'sound' Kendrick Frazier, editor of CSICOP's journal *Skeptical Enquirer*.

The sort of style adopted by members can be gauged from this observation on the organisation by Dr Peter Sturrock, professor of space science at Stanford University in California, writing in *New*

Scientist: 'In October 1632, Galileo was summoned to Rome to be examined by the inquisition for subtly but forcefully advocating the heliocentric theory. Today, a leading investigator of parapsychology, cryptozoology or UFO research may be politely invited to take part in a panel discussion at a CSICOP meeting.'[22]

This may not sound too intimidating until you learn of the behaviour of some CSICOP members. Professor Hansel, for example, believes that if he is able to conceive of any hypothetical way in which fraud *could* account for the results of a parapsychology experiment, then his 'rational reconstruction' constitutes proof that the experiment *was* faked. In one CSICOP publication, Hansel suggested a complicated means by which a famous series of ESP experiments at Duke University could have been faked. His scenario involved the subject secretly gaining entrance to a building, crawling through the attics and peeping through a hole in the ceiling while pretending to be guessing cards in another building. Hansel's suggestion was based on a personal visit he made to the university twenty years after the experiment, but he was unaware that the laboratory had been rebuilt since the experiment. When the original blueprints were produced to show that his scenario was impossible, Hansel ignored them and maintained his 'fraud' scenario in a reissue of his book by CSICOP's publishing house as recently as 1980.[23]

You might imagine that an organisation which calls itself a committee for scientific investigation, and whose declared aim is to carry out such research, would concern itself mainly or wholly with conducting scientific investigations, or at least with fostering or sponsoring such investigations by others. Surprisingly, CSICOP neither conducts nor sponsors any such investigations, and has not done so for fifteen years, since an unseemly squabble over its first and only foray into paranormal research.

Shortly after the organisation was first formed in 1976, it embarked on a study designed to replicate the work done by French researchers Michel and Françoise Gauquelin. The Gauquelins' work infuriated CSICOP (and continues to do so) because it provides an empirical basis for some of the claims of astrology – the French workers found, for example, an inexplicably regular relationship between the birth times of sports champions and the position of Mars.[24]

After a CSICOP team conducted its own investigation, Paul Kurtz announced in the *Skeptical Enquirer* that the Gauquelins' work could not be replicated and that there was no such correlation. This proved too much for CSICOP executive committee member Dennis Rawlins, who had acted as the study's statistician. Rawlins resigned from the committee saying that Kurtz and his fellow researchers had manipulated the data and had then engineered a cover-up when he attempted to bring the matter to the attention of other committee members. A number of other CSICOP members also resigned when the scandal was made public.[25]

The position today is that CSICOP no longer conducts any serious scientific investigations nor does it attempt to replicate those of others. Instead, it issues *ex cathedra* pronouncements through the medium of its own publications affirming its fundamentalist belief in orthodox science.

Rather more upmarket from CSICOP is Britain's COPUS, the Committee On the Public Understanding of Science. This organisation was founded in 1986 and is run jointly by Britain's most prestigious scientific institutions: The Royal Society, the Royal Institution and the British Association for the Advancement of Science.

Like CSICOP, COPUS is composed of self-elected guardians of the current paradigm. Unlike CSICOP, however, its position is not so much that the paradigm is in danger from irrational thinking and weird beliefs, but that we the public, because of our limited intellects and scientific ignorance, have failed to grasp the awesome nature and meaning of reductionist scientific discoveries. Consequently we persist in hanging on to outmoded concepts such as vitalist ideas in biology and the irrational belief that philosophy and history contain knowledge. By making heroic efforts in a missionary role the Committee hope that eventually they will convert the more enlightened among us to reductionism.

Amongst other activities COPUS makes grants of smallish amounts (around £2,000 or so) to finance worthy local scientific initiatives. Since 1987 it has handed out more than £400,000 on such ventures. The majority of these projects have clearly laudable aims that no rational person could complain at, such as making young people more aware of the importance of engineering; helping local museums build new scientific exhibits and so on. But you can still detect an unconscious egoistical undercurrent to some of these expenses, such as funding maps of cities showing where famous scientists lived. (How would members of COPUS react to public money being spent on maps showing where famous journalists lived?)

Although COPUS is a cut above CSICOP socially (its members are mainly scientific toffs such as its chairman Dr Lewis Wolpert, Professor of Biology at University College London) its intellectual level is the same. The Committee does not for example explain how, scientifically, it can know *in advance* that a reductionist explanation will ultimately answer all the questions of science. Indeed, such a belief looks perilously akin to scientific mysticism. Dr Wolpert revealed the kind of scientific attitude he believes is best for us, the public, to cultivate when he recently told the *Sunday Times,* "Open minds are empty minds."

The activities of organisations like CSICOP and COPUS rarely impinge on the public consciousness, but occasionally there are controversies that make the headlines. Probably the best-known former member of CSICOP is television conjuror James Randi, who became famous for his attempts to expose Uri Geller as a cheat.[26] Randi cheerfully describes himself as a conjuror who is skilled in misdirecting people, and he has employed his professional skills in attempts to discredit Geller and other 'psychics'. Randi claims that, in principle, he can duplicate by stage magic anything that Geller and other metal benders can do in the laboratory. In practice, though, he is able to replicate only some of those phenomena by conjuring tricks, and, unlike Geller and company, he does not submit himself to controlled laboratory conditions. Where he is unable to explain a particular phenomenon, he simply ignores it.

In the 1972 experiments at Stanford Research Institute, for example, there were four things that Geller did under controlled conditions, on videotape, that require explanation. The first was guessing the number on the face of a die shaken in a closed metal box, eight times. The second was a similar experiment in which he guessed the contents of sealed aluminium containers twelve times. The third was when he caused the read-out of an electronic scale to deflect in both the 'heavier' direction and the 'lighter' without touching it. The fourth was causing the needle of a gaussmeter to deflect without touching it. In his book 'exposing' Geller, Randi explained the first by suggesting that, in front of investigators Puthoff and Targ, Geller picked up the metal box, opened it slightly, peeped inside without them noticing and replaced the box. He repeated this process eight times, but is so skilled at misdirection that the investigators did not notice, and the actions were not recorded on the videotape monitor. Randi did not attempt to explain the sealed aluminium containers, the electronic scales or the gaussmeter. However, he concluded from this analysis that he had exposed Geller's SRI tests as fraudulent.

This depressing catalogue of cases inevitably provokes the question: precisely what is it that the guardians are guarding? Who or what are they guarding it from? And why have they selected themselves for this task?

I believe the answers to all these questions are contained in the preceding two chapters. They are guarding us, the community, from the awful effects of believing something that is irrational. (Meaning: they are guarding the current paradigm.) They are guarding us from the charlatans and fraudsters who play in an unscrupulous way on the fear and ignorance of ordinary members of the public, who, because we lack the necessary scientific knowledge and training, are unable to evaluate the nature of this threat for ourselves. (They know best what we should think.) And they have selected themselves because it is their duty to put their superior knowledge and expertise at the disposal of the community to protect us from the harmful effects of weird and unpredictable belief systems. (They have low tolerance of cognitive dissonance and are striving to reduce it to a minimum.)

Perhaps, though, science really does need groups like CSICOP. After all, if science does not defend itself, who will? Dr Richard Broughton is director of research at the Institute for Parapsychology in North Carolina. His view is that science's greatest strength is its ability to protect itself.

> Of course the real function of CSICOP is as an advocacy group to lobby for a particular point of view. Certainly the organisation is effective in this way, and few would deny that there is often a need to counter the public's credulity. But somewhere along the line

CSICOP abandoned the *objectively* critical approach and adopted a 'stop at any cost' approach toward any topic that it deems off-limits to science. Fortunately the scientific controversy over parapsychology will not be resolved at press conferences and in the media. Only in the appropriate professional forums can the give-and-take of science go on.[27]

'Science', observes Dr Broughton, 'is a marvelously self-correcting system. If there are errors or bad science, this will be weeded out in due course. Science does not need vigilantes to guard the gates.'

CHAPTER ELEVEN

······································

A Trout in the Milk

Some circumstantial evidence is very strong,
as when you find a trout in the milk.
HENRY THOREAU

One sunny afternoon in 1976, the normally sedate upper chamber of Parliament, the House of Lords, was startled from its post-lunch doze into sudden wakefulness by a question from one of its aristocratic members. When, demanded His Lordship the eighth Earl Clancarty, was the British government going to do something about the unidentified flying objects that were flying into and out of the Earth via the secret entrance situated at the North Pole? Wasn't it time the British people were told the truth about this scandalous cover up?[1]

Clancarty, better known as writer Brinsley le Poer Trench, had concrete evidence to back up his claim – photographs taken from space by a NASA satellite in 1968 which did, indeed, appear to show an unmistakable hole or crater at the Pole, several miles in diameter, exposing the dark interior of our planet. It was through this that the UFOs of his question made their entrances and exits.

Most people who consider themselves rational would probably say that the noble lord was completely off his rocker to believe such a thing. Yet Clancarty was basing his claim not on hearsay or imagination, but on a pile of both direct and circumstantial evidence that, taken in its entirety, amounts to a case for his claim – and moreover a case that must be answered. Merely to dismiss a carefully prepared body of evidence – however barmy it may appear – is to make the same mistake as the crank: to base a judgement on opinion rather than on empirical proof and logical argument. True scepticism demands more than mere rejection.

Common sense says that Lord Clancarty is wrong; yet his photographs and other data show he just could be right. The question is, how exactly are we to tell the difference between a real crank and someone who has stumbled across startling new knowledge, like Roentgen and X-rays? Or someone like the Wright brothers who really has achieved the seemingly impossible? Or a researcher who discovers a phenomenon for which the first evidence is weak and ambiguous, like the therapeutic value of aspirin for heart disease?

The hollow Earth idea is the kind of theory that makes most of us smile. Yet the evidence mustered by its adherents is far from weak. Or to be more exact, the secondary physical evidence can be interpreted to mean different things depending on what you already believe. Although this ambiguity can sow the seeds of doubt in a mind open in the literal sense, most people accept the orthodox view that the Earth is not hollow and that UFOs do not go into and out of Polar entrances. But what *exactly* is it that makes us so sure?

The common picture of the crank is, I suspect, that of the lone, obsessive character with wild eyes and an anorak, eager to buttonhole anyone who will listen, and espousing beliefs that most of us find at best slightly barmy and, at worst, barking mad. But it is not merely uneducated laymen and laywomen with untrained minds who become obsessed with crank beliefs, it is scientists as well.

Take the case of René-Prosper Blondlot, a physicist at the University of Nancy and distinguished member of the Académie des Sciences. In 1903, like many of his contemporaries, Blondlot was experimenting with the newly discovered X-rays. Through this he discovered hitherto unknown rays emitted by an incandescent filament. The rays would pass through aluminium but not through iron. In a very faintly lit room, said Blondlot, you could actually see the very slight increase in illumination that the rays caused when shone on to a piece of paper.

Blondlot announced these rays to the world in 1905. Since other physicists had already inconsiderately bagged the more obvious letters of the alphabet for their own forms of radiation (alpha, beta, gamma and even X) Blondlot decided to give his university of Nancy a little useful publicity by calling them 'N-rays'.[2]

As his experiments progressed, Blondlot discovered that N-rays had some truly extraordinary characteristics. He found he could store the rays by wrapping a brick in black paper and leaving it in the sunshine. Later when he recovered the brick and took it into the darkened laboratory, he could see a visible increase in illumination from the brick. Curiously, though, the amount of illumination always remained the same – he tried two bricks, three bricks, ten bricks, but he did not get any more illumination.

Blondlot next found that many things give off N-rays, including human beings. He found that radiant heat increased the effect of N-rays but that N-rays themselves were not identical with heat (infra-red) radiation because radiant heat will not pass through aluminium whereas N-rays will. Blondlot published a series of papers on his discoveries. Other researchers took up the same line of research and also published papers, about half of them supporting Blondlot's findings.

Eventually, the research attracted the attention of Dr R.W. Wood, a respected physicist from Johns Hopkins University, who decided to investigate Blondlot's phenomena at first hand and visited the Nancy laboratory. The set-up that Blondlot showed Wood consisted of a filament generating the N-rays which passed through a slit about 2 millimetres wide on to an aluminium prism and were refracted in the same way that a beam of visible light is refracted by a glass prism. Blondlot was able to measure the angle (and hence the refractive index) of several beams into which the N-rays were split by the aluminium prism and it was the making of these measurements that Blondlot demonstrated to the American, in his almost-dark laboratory.

Wood later described how his suspicions were first aroused by the precision with which Blondlot appeared to be making his measurements. The N-ray beam could not be less than 2 millimetres wide (the width of the slit through which it had to pass) yet the Frenchman claimed to be able to detect its position to within a tenth of a millimetre. When he questioned Blondlot on this point, the Frenchman told him, 'That's one of the fascinating things about the N-rays. They don't follow the ordinary laws of science that you ordinarily think of. You have to consider these things all by themselves.'[3]

Wood then asked Blondlot to repeat some of the measurements for him, but as the Frenchman began to do so, Wood surreptitiously picked up the aluminium prism and slipped it into his pocket, unnoticed in the darkness. Despite the fact that the prism was no longer there, Blondlot recorded the same refractive indices he had obtained earlier. Once Wood published this story, Blondlot was permanently discredited.

This example was recounted in some detail by Dr Irving Langmuir, a physical chemist who won the Nobel prize for chemistry in 1932 for his studies of molecular films. He worked from 1909 to 1950 in the US General Electric Company's research laboratory at Schenectady. In the course of his career he collected examples of what he called 'pathological science', cases of otherwise respected scientists deluding themselves into believing that they have discovered novel phenomena. Langmuir never published his work but in 1953 he held a colloquium on the subject at General Electric's Atomic Power Laboratory. In 1966, a recording of the lecture was discovered in the Library of Congress and a former colleague of Langmuir's, Robert N. Hall, later transcribed and edited the recording. The result was published in the magazine *Physics Today* in October 1989.[4]

The examples of 'pathological science' collected by Langmuir are worth examining in some detail because they provide many instructive clues to the nature of the self-delusion mechanism that affects people such as Blondlot. Equally, I believe, they illuminate the critical difficulty involved in separating the delusional from the novel.

The next example concerns an experiment carried out in 1930 at Columbia University by Professor Bergen Davis and his colleague Dr Arthur Barnes that attracted wide attention in the physics community. Davis and Barnes constructed an apparatus that generated alpha-particles (the nuclei of helium atoms) and electrons and they devised a way to send a stream of both types of particle through a glass tube towards a target. By changing the voltage on the tube they could adjust the speed of the electrons until it matched that of the alpha-particles and when they did, they believed that the alpha-particles captured the electrons as they flew along beside them to make whole atoms. This capture phenomenon was signalled by a dramatic decrease in the number of alpha-particles observed to arrive at the target.

Davis and Barnes could measure this decrease because they arranged that the particles coming out of the end of their apparatus would strike a screen and cause visible scintillations that could be seen through a microscope. Because each flash was quite small, they viewed the screen in the dark (a little like Blondlot looking at his N-rays).

If they set the apparatus so that no electrons at all were generated, all the alpha-particles were observed to strike the target screen, at the rate of about fifty counts per minute. When they turned the voltage up to a critical value, electrons were sent through the tube at the same speed as the alpha-particles, were captured, and only a few scintillations were visible.

Langmuir and his colleague, Clarence Hewlett, visited Columbia University to see the Davis–Barnes experiment for themselves. Langmuir describes what happened next.

> We sat in the dark room for half an hour to get our eyes adapted to the darkness so that we could count scintillations. I said I would first like to see these scintillations with the field on and with the field off. So I looked in and I counted about 50 or 60. Hewlett counted 70, and I counted somewhat lower. On the other hand we both agreed substantially. What we found was this: these scintillations were quite bright with your eyes adapted, and there was no trouble at all counting when the alpha particles struck the screen. They came along at the rate of about one per second. When you put on a magnetic field and deflected them out, the count came down to about 17, which was still a pretty high percentage – about 25% background. Barnes was sitting with us, and he said that's probably radioactive contamination of the screen. Then Barnes counted and he got 230 on the first count and about 200 on the next, and when he put on the field, the count went down to about 25. Well, Hewlett and I didn't know what that meant, but we couldn't see 230. Later, we understood the reason . . .

Langmuir then played what he himself called 'a dirty trick'. The various voltages being applied to the apparatus were under the control of a laboratory assistant called Hull. The number of scintillations observed, said Davis and Barnes, depended on the exact voltage applied. They further claimed that the effects were sensitive to even a one-hundredth of a volt change. Langmuir now wrote out a list of ten test voltages, some of which were zero, and passed this to Hull, thinking to catch Barnes out with the zero readings. The trick failed to work at first, because when Hull was not reading a voltage he sat back in his chair, away from the instrument, thus giving Barnes a subconscious clue that there was nothing to measure, and he registered a nil result.

Langmuir explained: 'So I whispered [to Hull] "don't let him know you're not reading" and I asked him to change the voltage from 325 down to 320 so he'd have something to regulate. I said, "regulate it just as carefully as if you were sitting on a peak [of electron capture]". So he played the part from that time on, and from that time on Barnes's readings had nothing whatever to do with the voltages that were applied. Whether the voltage was at one value or another didn't make the slightest difference. After that he took 12 readings, of which about half were right and the other half wrong, which was about what you would expect out of two sets of values.'

Langmuir confronted Davis with his findings. 'He couldn't believe a word of it. "It absolutely can't be", he said. "Look at the way we found those peaks before we knew anything about the Bohr theory. We took those values and calculated them and they checked exactly." . . . He was so sure from the whole history of the thing that it was utterly impossible that there never had been any measurements at all that he just wouldn't believe it.'

Langmuir also describes what he felt sure was the cause of Davis and Barnes's delusion. '[Barnes] was counting hallucinations, which I find is common among people who work with scintillations if they count for too long. Barnes counted for six hours a day and it never fatigued him. Of course it didn't fatigue him, because it was all made up out of his head. He told us that you mustn't count the bright particles. He had a beautiful reason why you mustn't pay any attention to the bright flashes. When Hewlett tried to check his data [Barnes] said: "why, you must be counting those bright flashes. Those things are only due to radioactive contamination or something else." He had a reason for rejecting the very essence of the thing that was important.'

Langmuir wrote to Danish physicist Nils Bohr to 'head off' any further experimentation. A year and a half later, Davis and Barnes published a short letter in *Physical Review*[5] saying that they had been unable to reproduce the effect. 'The results reported in the earlier paper,' they said, 'depended on observations made by counting scintillations visually. The scintillations produced by alpha particles on a zinc sulphide screen are a threshold phenomenon. It is possible that the number of counts may be influenced by external suggestion or autosuggestion to the observer.'

Langmuir observed that, 'to me [it is] extremely interesting that men, perfectly honest, enthusiastic over their work, can so completely fool themselves. Now what was it about that work that made it so easy for them to do that?'

Langmuir's next example was one described earlier, that of mitogenic rays associated with living organisms described by Professor Alexander Gurwitsch of the First State University of Moscow in 1923 (see Chapter 5). The rays were believed to be a form of ultraviolet light emitted by all living cells because they would pass through a quartz screen but not through glass. Gurwitsch and others found that they could just detect mitogenic rays on a photographic plate.

Of these experiments, Langmuir says that 'if you looked over the photographic plates that showed this ultraviolet light you found that the amount of light was not much bigger than the natural particles of the photographic plate so that people could have different opinions as to whether it did or did not show this effect. The result was that less than half of the people who tried to repeat these experiments got any confirmation of it . . .'

In the example of mitogenic rays, Langmuir seems to have shifted his ground somewhat. He has moved from experiments that were difficult or impossible to confirm to those that are confirmed by only 'less than half' those who try to replicate them. He has also moved from phenomena that are wholly imaginary to those that are difficult to discern or discriminate, and where it is difficult or impossible to say on the data presented whether they were real or not.

When he was delivering his speech in 1953, Langmuir was not to know that two decades later, in 1972, Gurwitsch's experiments would be repeated and strikingly confirmed using modern instrumentation by S.P. Shchurin and a team from the Institute of Clinical and Experimental Medicine in Novosibirsk, Russia.[6] Langmuir was partly basing his conclusion on the statement of the American Association for the Advancement of Science that Gurwitsch's experimental results were delusory and mitogenic rays all in his mind.

Langmuir's final example is even more striking than those given earlier. It concerns research conducted by Professor Fred Allison of the Alabama Polytechnic Institute and published in 1927. As a result of Allison's work, scores of papers were published in journals such as *Physical Review* and *Journal of the American Chemical Society*, and Allison and his co-workers even claimed to have discovered new isotopes and new elements. Indeed, as we will see later, there is a good case to be made for saying that Allison was instrumental in the first confirmation of the existence of tritium, the rarest isotope of hydrogen.

The apparatus built by Allison made use of an effect first noticed by Michael Faraday: the rotation of a beam of polarised light by a magnetic field. If you shine a beam of polarised light through a liquid, and then turn on a magnetic field round the liquid, the plane of polarisation of the light is rotated by the magnetism. You can see this effect easily because the light beam will appear to get lighter or darker as it rotates.

The apparatus Allison and his co-workers built had two glass tubes in line that could be filled with liquids, with coils of wire wound round them in a circuit. The light source was an electric spark that also sent a current through the coils. They could visually observe the amount of rotation due to the Faraday effect by looking down the second glass tube and rotating it until it just compensated for the amount of rotation caused in the first tube. However, Allison also discovered that the amount of rotation depended on a second factor: the composition of the liquids inside the glass tubes. If he used plain water in the second tube he got one reading. If he dissolved some salt in the water he got another reading.

Allison made use of this ability to discriminate between solutions as an analytical tool. For instance, ethyl alcohol produced one characteristic measurement, while acetic acid produced another. But when he put in ethyl acetate (a compound of both chemicals) there were two characteristic measurement peaks. This meant that he could analyse compounds using the 'magneto-optic' apparatus.

A further important point about this ability to discriminate solutions was that it appeared to be amazingly sensitive. He got the characteristic measurement peaks from the compounds in solution even if that solution was as weak as a 10^{-8} molar solution, which is something like a pinch of salt in a bathful of water.

The most important discovery came when Allison realised that his apparatus was capable of discriminating between different isotopes of the same element. This was remarkable because discriminating isotopes by chemical methods is usually a very complex and time-consuming process, while Allison's apparatus was relatively simple. Using his apparatus, Allison announced that there were sixteen isotopes of lead. A little later, he made an even more important discovery: that the apparatus could also detect entirely new elements. By the 1920s, the modern periodic table of elements had been drawn up, but chemists had not yet been able to identify physically some of the elements that it predicted should exist. Using his effect, Allison now came up with two new elements which he named Alabamine and Virginium.

The end of the Allison story is curiously inconclusive and ambiguous. Langmuir left his audience in no doubt that he personally believed the 'Allison effect' was an example of pathological science; that it existed only in the mind of its discoverer and his colleagues. Yet Langmuir himself tells the following story, concerning two of his friends and colleagues, Gilbert Lewis, professor of physical chemistry at Berkeley, and Wendell Latimer, head of the chemistry department at the University of California.

Langmuir records Latimer as saying, 'There's something funny about this Allison effect, how they can detect isotopes. I think I'll go down and see Allison, to Alabama, and see what there is in it. I'd like to use some of these methods.'

Latimer's main interest at that time, says Langmuir, was the talk there was in physical chemistry circles about the possible existence of traces of hydrogen of atomic weight 3 (today called tritium) for which there was some spectroscopic evidence. Latimer shared these thoughts with Gilbert Lewis, who is said to have replied, 'I'll bet you ten dollars that you find there's nothing in it.'

Latimer visited Allison in Alabama and stayed three weeks studying his methods. He returned to the University of California, built a duplicate apparatus and got it working so well that Lewis paid him the ten dollars. He identified tritium using Allison's magneto-optic method and published a short paper in *Physical Review* in 1933 announcing the detection of the isotope of hydrogen of atomic weight 3.[7] Curiously, however, most textbooks credit the discovery to Ernest Rutherford and colleagues in the following year.

Langmuir then recalled that around this time there was a meeting of the American Chemical Society at which there was much discussion about whether to accept any more papers on the Allison apparatus for its *Journal*. The decision, says Langmuir, was against accepting any more papers, a ruling that he says was also adopted by *Physical Review*.

Despite adopting this policy, however, both journals did publish other papers on the Allison effect, both of them by groups who felt that a perfectly valid effect was being marginalised and suppressed. The paper in the *Journal of the American Chemical Society* in 1932, by J.L. McGhee and Margaret Lawrentz, said that, 'In December 1930 one of us (McGhee) handed out by number to Prof. Allison twelve (to him) unknowns which were tested by him and checked by two assistants 100 per cent correctly in three hours.'[8]

The paper by T.R. Ball of Missouri's Washington University published in *Physical Review* in 1935 gave a very detailed review of the magneto-optical method and included a statistical study of 1,698 readings made over three years by five different observers. It included the correct identification of unknown substances in several blind tests where the odds were 34 to 1 against selection by chance alone.[9]

However Langmuir concludes that the Allison effect is nevertheless an example of pathological science because after discovering tritium, Latimer told him: 'You know, I don't know what was wrong with me at that time. After I published that paper I never could repeat the experiment again. I haven't the least idea why. But those results were wonderful. I showed them to G.N. Lewis and we both agreed that it was all right. They were clean-cut. I checked them myself every way I knew how to. I don't know what else I could have done, but later on I just couldn't ever do it again.'

From his list of examples, Langmuir suggested the existence of a generalised pathological approach to research that can afflict perfectly honest scientists, and he drew up a list of tell-tale signs by which such research can be recognised.

'The Davis–Barnes experiment and the N-rays and the mitogenic rays all have things in common. These are cases where there is no dishonesty involved but where people are tricked into false results by a lack of understanding about what human beings can do to themselves in the way of being led astray by subjective effects, wishful thinking or threshold interactions. These are examples of pathological science. These are things that attracted a great deal of attention. Usually hundreds of papers have been published on them. Sometimes they have lasted for 15 or 20 years and then they gradually have died away.'

The symptoms of pathological science, according to Langmuir, are as follows:

- The maximum effect that is observed is produced by a causative agent of barely detectable intensity. Moreover, the intensity cannot be increased by multiplying the source: ten bricks still only yielded the same number of N-rays as one brick. This, says Langmuir, is to make it easier to fool yourself.
- The effect is near the threshold of visibility or the threshold of any other sense used to detect it. Alternatively, many, many measurements are needed because of the very low statistical significance of the results. This enables people to find plausible reasons for rejecting data that does not fit. 'Davis and Barnes were doing that right along', says Langmuir. 'If things were doubtful at all, why, they would discard them depending on whether or not they fit the theory.'
- There are claims of great accuracy, great sensitivity, or great specificity. This was particularly true of the Allison effect, says Langmuir.
- Fantastic theories contrary to experience are suggested.
- Criticisms are met by *ad hoc* excuses thought up on the spur of the moment.
- The ratio of supporters to critics rises up to somewhere near 50 per cent and then falls gradually to oblivion.

In the example of Allison, Langmuir seems to have shifted his ground yet again in defining what constitutes pathological science. It is easy to see why his suspicions were aroused in the light of Latimer's remarks to him. Equally one can see that Allison's effect conforms in many respects to the definition of classic pathological research: it depends on visual observation of a critical measurement; it claims to function without any variation at fantastically high dilutions; and Allison came up with sixteen isotopes of lead while orthodox chemistry recognises only three isotopes (although acknowledging that discriminating the atomic weights of these isotopes is very difficult). There is also the issue of the so-called new elements Virginium and Alabamine (the latter ostentatiously named after the discoverer's own state poly-technic, rather like Blondlot's N-rays), which do not appear anywhere in the periodic table today.

Yet it seems to me that in setting out quite properly to nail abuse of the scientific method, Langmuir has unconsciously drifted across the line that separates his definition of pathological science from the foggy zone out of which real discovery emerges, faint and fuzzy round the edges. How was Allison able to identify twelve unknowns in three hours with 100 per cent success, if it was all in his mind? How were Ball and his colleagues at Washington University able to repeat similar blind tests if the effect was imaginary? How did Latimer dis-cover tritium in the first place?

As Langmuir himself said of McGhee's paper (the *last* such paper published by the American Chemical Society) 'You'd think that would be the beginning, not the end.' You would, indeed. Except that both the *AMC Journal* and *Physical Review* took a policy decision *not* to publish anything further on the Allison effect, and prominent physical chemists like Langmuir were discouraging any further research. So far as I know, no one has recorded Wendell Latimer's reaction to not being credited with the discovery of tritium, but seeing the credit go a full year later to Britain's Ernest Rutherford.

In setting out to evaluate Langmuir's hypothesis of pathological science, I want to attempt to apply his diagnostic techniques to a rather different kind of experiment: one drawn from the realms of physics, but one that has assumed a position central to modern science and about which there is absolutely no doubt in orthodox circles. The experiment I have in mind is that first carried out by Dr Robert Millikan, director of physics at the California Institute of Technology and winner of the Nobel prize in 1923 for measuring the charge of the electron.

This experiment was crucial to the development of twentieth-century atomic physics. It established that all electrons have the same unit charge, and that there are never fractions of this charge found in nature. And for the first time, Millikan accurately quantified that minutely small charge. The experiment he devised to enable him to measure something so small was brilliantly ingenious.

He made tiny droplets of oil fall through a hole in the lid of a box with transparent sides and he observed through a microscope the time they took to fall. From this he was able to deduce their size (using an existing formula). He then switched on a voltage between the top and bottom plates of the box. From the effect on the rate of fall he was able to calculate any electric charge on the drop. Sometimes this charge would abruptly change, but by a constant amount, and Millikan reasoned that this was because it had either lost or captured an electron. By performing this experiment over and over, searching carefully through all his measurements of many thousands of drops, Millikan was able to find the number that represented the charge of a single electron (he found it to be 1.6×10^{-19} coulombs). This measurement provided the basic scale parameter for the whole of atomic physics.

What is particularly interesting is that Millikan's laboratory notebooks have survived and so we can gain an insight into his thought processes in the very moments that he was conducting his famous experiment, a little like paying the sort of personal visit favoured by Irving Langmuir. What those notes reveal, however, is just as suspicious as anything that Langmuir turned up in his investigations.

After one run of measurements, Millikan noted, 'This is almost exactly right!', while after another less satisfactory trial he noted down, 'very low – something wrong'. How did Millikan know in advance what was the 'right' result and what was 'wrong'? Does this mean that Millikan was acting in some way dishonestly – setting out to prove a preconceived idea? No, it merely shows that scientific research is a complex business that sometimes depends on intuition as much as on deduction (and that it is foolish to expect scientists not to theorise in advance of the data).[10]

Millikan's experiment has been repeated many times since. Indeed it is frequently carried out by school and university students because it is in essence so simple. Yet anyone who has ever repeated Millikan's experiment knows that you never, ever get exact results, and that you frequently do apparently find fractions of a unit charge. These findings were rejected by Millikan as due to experimental error, and are still rejected today because the atomic theory is so powerful and because Millikan's experiment is crucial to that theory.

Some modern scientists believe that there *are* fractional atomic charges. Professor Bill Fairbank of California's Stanford University has claimed to have measured particles with one-third of the charge of the electron in experiments which have been repeated over a five-year period with increasing accuracy, using apparatus far more sophistic- ated than that used by Millikan in 1916. Against this, Dr Peter Smith of the UK's Rutherford-Appleton laboratory has constructed similarly sophisticated apparatus which he claims does not replicate Fairbank's results.

Regardless of the outcome of this modern controversy, how does Millikan's original experiment stand up to Irving Langmuir's criteria? The answer is that it fails on practically every point of comparison. It is a threshold effect involving visual observation of microscopically small oil particles. The magnitudes involved are unimaginably small. Large numbers of trials have to be carried out and statistical results closely examined for trends. Some data is rejected as being 'wrong' because it does not fit the theory, while other data is accepted as being 'almost exactly right'. Fantastic accuracy is claimed for the result. *Ad hoc* excuses are offered for the existence of anomalous data ('experimental error').

So why did Langmuir not include Millikan's experiment in his catalogue of pathological science? What *scientifically* distinguished Millikan's work from Allison's? The honest answer is that nothing so distinguished it except the final criterion that more than 50 per cent of scientists supported his results and have continued to do so. In the final analysis it was a matter of *acceptance*, not a matter of *evidence*.

Langmuir's lecture on 'pathological science' was published in the US magazine *Physics Today* in October 1989, a few months after Fleischmann and Pons had announced their discovery of cold fusion and shortly after a speaker at the American Physical Society's annual meeting in Baltimore said that physicists were 'Suffering from the

incompetence and perhaps delusion of doctors Pons and Fleischmann.'
Although *Physics Today* published Langmuir's lecture without direct
comment, even the most unworldly reader could see who it was aimed
at, given the furore that the two scientists had created. In the United
Kingdom, science journalists were nothing like as reticent about
making the connection and accused Fleischmann and Pons openly.
Writing in the *Daily Telegraph* Steve Connor asked, 'how two
respected chemists could apparently make such a blunder'? The
answer, supplied Connor helpfully, was that they were victims of
Langmuir's pathological science.[11]

At the height of the Velikovsky affair described in the previous
chapter Dr Laurence Lafleur, associate professor of philosophy at
Florida State University, wrote an article in *Scientific Monthly* also
setting out to define the criteria by which we will be able to recognise a
crank from a real scientist.[12] Lafleur gave seven criteria, as follows:

1 Is the proposer of the hypothesis aware of the theory he proposes
to supersede?

2 Is the new hypothesis in accord with currently held theories in the
field of the hypothesis, or, if not, is there adequate reason for
making changes, reasons of weight at least equal to the weight of
evidence for the existing theories?

3 Is the new hypothesis in accord with the currently held theories in
other fields? If not is the proposer aware that he is challenging an
established body of knowledge, and does he have sufficient evidence
to make such a challenge reasonable?

4 In every case where the new hypothesis is in contradiction with an
established theory, does the hypothesis include or imply a suitable
substitute?

5 Does the new hypothesis fit in with existing theories in all fields,
or with substitutes proposed for them, to form a world view of an
adequacy equivalent to that of the currently accepted one?

6 If the new hypothesis is at variance with theories capable of
prediction or of mathematical accuracy, is the new theory itself
capable of such prediction or mathematical accuracy?

7 Does the proposer show a predisposition to accept minority
opinions, to quote individual opinions opposed to current views,
and to overemphasise the admitted fallibility of science?

Lafleur goes on to add, rather disingenuously, that, 'It is not our primary purpose to examine the merits of Velikovsky, but in defence of his critics it is necessary to point out that he qualifies as a crank by almost every one of these tests, perhaps by every one.'

It is rather difficult to tell now, from a distance of more than forty years, whether Professor Lafleur actually expected his remarks to be taken seriously outside of the context of the Velikovsky affair. For it is hard to see how any of the seven criteria above could be anything other than remarks addressed directly to Velikovsky. It is not until we get to proposition 6, that we even hear about the quality of the evidence that supports the new hypothesis: that is to say, the scientific issues themselves. The first five propositions are concerned almost entirely with how far the new idea offends against the beliefs of the elders and betters of science and whether such disgraceful impertinence can be justified. The seventh is even more bizarre: the fewer the people who agree with you, the more likely you are to be wrong. Again we have a criterion which ignores the scientific evidence and dwells instead on the crucial importance of maintaining a scientific consensus, whatever the facts may say. Again it is acceptance that counts, not evidence.

A better idea of Lafleur's true position on Velikovsky is gained from the introductory paragraphs of his article:

> the general public as represented by the editors and readers of *Harper's* has failed to grasp the reasons for the scientific rejection of Velikovsky's hypothesis, and many of them may therefore be led to think of scientists as a dogmatic crew, blindly maintaining their own unverified doctrines; intolerant of opposition, and suppressing it by denying free expression to their adversaries. The scientists, in general, have not been aware of the enormities attributed to them. They have not realised that the tempest is over something more than the purely scientific question; therefore, since they know that all other scientists agree with them in the rejection of Velikovsky's hypothesis, they are inclined to consider the question closed and turn their attention to less depressing matters.

The 'us and them' attitude; the patronising assumption that non-scientists' dissent is caused by their ignorance; the casual distortions ('all other scientists agree . . .') are still attitudes that can be found in science today, although it would be a great deal harder these days to find any scientist or philosopher willing to sign his name to them so publicly.

One present-day scientist who is concerned at the perception of anomalous phenomena as 'heretical' is Dr Peter Sturrock, professor of space science at Stanford University in California, who together with like-minded scientific colleagues, such as Professor Laurence Frederick of the University of Virginia, founded the Society for Scientific Explorations in 1982.

Sturrock points out that 'Sometimes, the nature, and even the reality, of a phenomenon may be obscure. If the case rests on only a few reports scattered in dusty, inaccessible journals, they cause no problem: they are simply ignored. If, however, the reports are so frequent that the topic cannot be ignored by the scientific community, scientists may become emotional and confused. Is it possible that the difficulty with "anomalous phenomena" is that scientists may perceive them as heresies?'[13]

Sturrock says that within his own field of astronomy some new claims are regarded as respectable and others as heretical. Quasars and, later, pulsars were truly anomalous objects but astronomers readily accepted them.

> On the other hand some astronomers, notably Halton Arp, now at the Max Planck Institute for Physics and Astronomy in Munich, claimed to have evidence that the red shift of quasars cannot be due entirely to processes that are already known. There has been great resistance to this claim, and Arp says that this resistance led to political repercussions that forced him to leave his observatory in the US for a more congenial place of business.

Sometimes, says Sturrock, anomalies deserve to sink without trace, and he cites Irving Langmuir's symposium on pathological science. But,

> the problems of evaluating anomalous phenomena would be solved more rapidly if it were only a question of applying the right scientific methodology. Unfortunately, however, the interested community may be organised into advocacies, arguing powerfully either for or against the reality and the importance of a particular

phenomenon. There are clearly collective processes at work. That is to say, social and political factors play a role in determining how the scientific community responds to anomalous phenomena.

Sturrock offers the following as guidelines to those willing to research anomalous phenomena;

- In studying any phenomenon, face up to the strongest evidence you can find, even if it is in conflict with current orthodoxies.
- Go to the original sources for your data. Do not trust secondary sources.
- Deal with 'degrees of belief', which can be conveniently characterised by probabilities. It is important to avoid assigning probability $P = 0$ (complete disbelief) or $P = 1$ (complete certainty) to any proposition since, if you adopt either of these values, that value can never be changed no matter how much evidence you subsequently receive.
- Focus on evidence and testing.
- Subdivide the work into categories so different people take on different tasks.
- Where possible work in teams; first because a combination of expertise may be required, and secondly, because a team is more likely to be self-correcting than someone working alone.
- In theoretical analyses, list all assumptions. This seems a simple, innocuous request, yet it will not always be easy to put into effect.

In one sense, it is refreshing to read a physical scientist advocating such a rigorously open-minded approach to anomalous phenomena. Sturrock's guidelines contrast starkly with Lafleur's dogmatic 'if it's different, it's wrong' approach. Yet in another sense, it seems to me almost incredible that three centuries after Galileo, some professional research scientists should have to receive counselling on how to be open-minded in research.

Most of the examples of crank beliefs cited earlier in this chapter have quite a high entertainment value and are capable of providing us with that most delightful of human pleasures: a laugh at someone else's expense. But the laughter is apt to obscure what is perhaps the most important lesson in all this: that a crank is not only one who, through self-delusion rather than evidence, believes a theory to be true when it is actually false. A crank is also one who, through self-delusion rather than evidence, believes a theory false when it is actually true. Thus, it is not only Lord Clancarty and believers in a hollow Earth who are cranks; it is also Lord Kelvin, who thought that X-rays were a hoax, and the editor of *Nature*, who said that cold fusion is 'licensed magic' and a waste of time.

....

CHAPTER TWELVE

Too Insensitive to Confirm

One should never believe any experiment until
it has been confirmed by theory.
Sir Arthur Eddington

There is something almost unbearably poignant about the image of
Orville Wright flying over Huffman's Prairie outside his home town
of Dayton, Ohio, while for five long years the tiny figures below
denied that he had ever left the ground.

Orville continued to circle through the clouds until President
Theodore Roosevelt ordered a public trial at Fort Myers in 1908 and
the Wrights were finally able to prove their claim to be the first to fly
in a powered, heavier-than-air machine. Until then their claims were
derided and the two young men dismissed as hoaxers by the *New York
Herald*, the *Scientific American* and the US Army.

As well as being yet another example to add to the dispiriting
catalogue of scientific prejudice and intolerance, the case of the
Wright brothers has a number of interesting features. The first aircraft
they built and flew at Kitty Hawk, North Carolina – the prototype of
the machine that circled over Huffman Field – only just managed to
get off the ground. Although it officially made the world's first
powered and controlled flight from level ground without take-off
assistance, the Wright Flyer's first performance was marginal to say
the least. It was airborne for only 12 seconds and travelled only 120
feet at an 'altitude' about shoulder high – scarcely more than one
could decently throw a glider.[1] The power of its home-made motor
was insignificant compared with the engine that powers a modern
aircraft. No one witnessing this first flight would have realised that
thousands of man-hours of theoretical design effort and experiment

had gone into its construction, including the first ever wind-tunnel tests to develop a controllable aerofoil wing.

In many respects, these first attempts to fly conform to Irving Langmuir's primary symptom of 'pathological science' in that they were barely measurable threshold phenomena that existed principally in the imagination of their originators.

It is only in the light of subsequent developments – which saw true controlled flight over substantial distances – that you can see this first attempt at Kitty Hawk as the historic first flight. Had this first effort not been taken up and developed then the North Carolina trials could equally well have been perceived as an experiment confirming the view of the Army and Professor Newcomb, that such powered flight is impossible, just as Robert Millikan's experiments could have been taken to mean there *are* particles with fractional charges.

Powered flight very nearly did not get off the ground at all; just like cold fusion almost a century later. Only determination and persistence by their discoverers compelled the community's acceptance – not scientific evidence, nor rational debate.

Examples like this raise the question: how many other important inventions and discoveries were not so fortunate? How many pioneers lacked the determination or the resources to compel acceptance or were simply unable to attract sufficient attention? Or again: how many inventions and discoveries were so marginal in their prototype or experimental form that they seemed to confirm the skeptics' viewpoint – rather like cold fusion being, in the memorable words of MIT, 'too insensitive to confirm'?

The orthodox rationalist view, as exemplified by Irving Langmuir, is that natural phenomena either occur or they do not occur. Facts are facts and if a scientific theory is valid then it can be experimentally verified under appropriate test conditions. The problem with this view is demonstrated by the many examples given in previous chapters. In their earliest manifestations, some novel phenomena appear weak, ill-defined and hardly more significant than the background noise against which they stand out. The Wrights struggled to stay airborne for a few seconds as Marconi strained to hear the first wireless radio transmission, or Joseph Niepce peered to discern the faint image of the first photograph. All were threshold phenomena initially. Equally, these embryonic phenomena are feeble and vulnerable to early extinction, just as living embryos are precarious and defenceless. The nailed boot of scientific derision alone may be enough to trample the life out of such delicate seedlings.

In these circumstances, the requirement that valid experiments must be repeatable at will takes on a specially crucial role in the acceptance or dismissal of any novel phenomenon. If an experimental result is significant only to one who has eyes to see that significance, is it any wonder that Nina Kulagina's ability to cause burn marks on researchers' arms is ignored, or that the boiling heat of the cold fusion jar is dismissed?

The wall of indifference and incredulity with which orthodox science greeted the Wright brothers' discovery of how to fly was also a 'first' in another sense that has become important in this century: they were among the earliest of a long and growing line of innovators who were not professional scientists. This trend away from professional science has become very pronounced in the second half of the century. It is the reversal of a historical trend that began when modern physics was born in seventeenth-century Italy.

Galileo replaced the largely Aristotelian scientific view, based on logic and verbal reasoning, with the modern mathematically based treatment of mechanics that has sustained physics for three centuries and that led to the reductionist mechanistic world view that dominated science throughout the eighteenth and nineteenth centuries. However, Galileo's replacement of the Aristotelian paradigm also had a less obvious but equally important consequence. Before Galileo, as science was based on verbal reasoning, anyone who took the trouble to become educated could understand the current scientific paradigm. This was no simple task in the Middle Ages to be sure, when the means of education were sparse and not easily accessible. But, in principle, any thinking person could discuss the current state of science and debate the merits of this or that idea with his or her fellow natural philosophers.

After Galileo, it was no longer possible for someone with an ordinary education in the humanities to study or debate the scientific issues – it became essential to have specialist knowledge of mathematics to do so. Science ceased to be the province of all educated people and became the property of a few specialists who handed down their conclusions to the rest of the community in much the same way that judicial decisions were handed down from the bench or papal bulls issued on religious matters. It was in this period that Europe's learned scientific societies were founded and it is from this time that the status of scientists as voices of authority dates.

The process might have continued indefinitely had it not begun to be reversed by the mechanical revolution of the steam age. By 1850, every railway stoker had a table of boiler pressures in his back pocket and by 1900 every bright boy or girl knew as much mathematics as Galileo, thanks to universal primary and secondary education and the appetite of industry for skilled workers. Ironically, it was the lads from the bicycle shop who took their technical education and made it yield up to them the secrets of flight: the secrets that had eluded the most intellectually gifted in the land. The rejection of their discovery was also a rejection of the idea that workmen with dirty overalls could lead mankind into a new scientific era.

But the scientific revolution initiated by Galilean science had other far-reaching consequences for the acceptance of new ideas. Before Galileo, most thinking people – including scientists – believed that a heavy weight drops to the Earth faster than a light weight. This view seems 'reasonable' and is consonant with intuition. Galileo's greatness consists in his daring to question and overturn a 'reasonable' and intuitive belief. He showed that the acceleration due to gravity is the same for every object no matter how massive – a 100 ton weight will fall at the same rate as a 1 ounce weight (neglecting air resistance and other factors). This view is both 'unreasonable' and counter-intuitive – yet it is unimpeachably correct by every scientific test.

Much of western science is constructed in this way. It is a continuing process of replacing intuitive, 'reasonable' but false beliefs with empirically tested theories – whether they seem reasonable or not, or run counter to our expectations. Almost all scientists would accept some such rough definition of how science proceeds. Yet how many would accept its necessary corollary – that we must inevitably still be living with some scientific beliefs that *seem reasonable yet are untrue*, simply because they have not yet been challenged by their Galileo?

To most dedicated professional scientists it is hard to have to admit that the progress of science should be determined to any extent by such a purely fortuitous factor. Yet our being prisoners of fortune should come as no surprise to anyone who studies how we, as a community, decide that a new idea is 'true' and acceptable. The conventional scientific view is that to be acceptable a new hypothesis has to pass the empirical test. If it survives the experiment, it is true. The 'black box' either works or it doesn't. But is this really how we evaluate the truth of scientific ideas?

Newton's theory of gravitation was accepted as true before Einstein replaced it. Does this mean that all physicists before Einstein were wrong? Of course not. The European Enlightenment and the Industrial Revolution were built upon Newtonian mechanics. The artilleryman's shell still falls on its predicted target with lethal precision, even though Newton's physics is cosmically inexact, and the gunner knows nothing of relativity. The replacement of Newton by Einstein is merely an example of the normal course of science proceeding from error to correction to fresh errors, which are often illuminating in themselves. The search by physicists at the end of the nineteenth century for the 'luminiferous ether' proved to be highly productive and eventually led Einstein to relativity theory.

> It is a layman's illusion [wrote Sir Peter Medawar] that in science we caper from pinnacle to pinnacle of achievement and that we exercise a Method which preserves us from error. Indeed we do not; our way of going about things takes it for granted that we guess less often right than wrong, but at the same time ensures that we need not persist in error if we earnestly and honestly endeavour not to do so.[2]

But if science is always in a state of 'controlled wrongness', so to speak, a state in which its current imperfections are awaiting improvement in the next paradigm shift, then at what point can we say that we know for sure a new idea works or is true using the scientific method as our yardstick? Put this way, science seems unlikely ever to provide us with the certainty that an entirely new idea is true or false except when our knowledge is complete – and there are no more new ideas.

Perhaps if a theoretical test eludes us there is still the practical test? We don't have to know *how* something works to prove that it does so – an aerofoil wing, for instance – we can try it empirically.

Disappointingly, the practical approach is no more enlightening. All complex mechanical and electrical systems are unstable in most configurations and continue to function only because they are in some sense 'nursed along' or kept in tune by skilled operators or maintenance staff. This is true of all motor cars, all aircraft and all computers, for example, all of which periodically fail and have to be repaired. The implication of this fact is that we can only know whether a complex system (like the Wrights' aircraft) works when it is in running order; but we can't know for sure whether it is in running order if it is entirely novel and operates on principles counter to the current paradigm. Imagine trying to repair a motor car with a broken fuel pump if you don't know what a motor car is or even what petrol is for.

The aero engine used by the Wright brothers was pathetic by the standards of the Rolls-Royce jet that powers Concorde and was only just powerful enough to get the pioneer aviators off the ground. It only just worked. Perhaps, then, a 'black box' which diagnoses illness only just works and is awaiting development like the aero engine. Or perhaps faint signals like Alexander Gurwitsch's mitogenic rays are merely awaiting amplification by suitable techniques. Or the signal detected by the dowser is undetectable by our current instruments. The important point here is not whether these conjectures are correct, but that there is *nothing* in the scientific method that will reveal for certain whether this is the case or not. It is only later, after a successful amplification of the embryonic discovery, that we may apply the scientific method with the certainty that we are neither fooling ourselves nor being fooled by a threshold phenomenon.

If we are denied a clear-cut answer from the empirical test, then what else is left to be the arbiter of what is acceptable and true? From the examples in this book, many readers may well draw the conclusion that the only such arbiter is fate, fortune or lady luck. Those inventions and discoveries that have found a place in our lives are not carefully selected by stringent tests in the laboratory and the economic rigours of the market as the best available, or even as meeting our specific needs. On the contrary, the science and technology we have is more or less the product of chance.

It would have been perfectly possible technically, for instance, to have constructed a high-definition electronic scanning television system while Queen Victoria was still on the throne in the late 1890s. Even the scientist to perform the task was available in the form of the young Ernest Rutherford who was attracted to Marconi's experiments with wireless. But Rutherford was advised by his mentor Lord Kelvin not to have anything to do with radio since it had no practical application, but to study radioactivity as a more promising field.

The shift to amateur discovery typified by the Wright brothers has led to a revolutionary change in scientific research in recent decades, yet one that has scarcely been noticed. It is exemplified in its most spectacular form by Michaela and Augusto Odone, the American parents whose son, Lorenzo, contracted the hereditary disease adreno-leukodystrophy (ALD) in 1983. Doctors told the Odones that ALD attacks the body's myelin, the fatty tissue that sheaths nerve fibres, and is invariably fatal. In two years, they said, Lorenzo would be dead. Unwilling to accept this verdict and simply watch their son die, the Odones – with no scientific training – undertook their own bio-chemical research and discovered a combination of oils that arrested the course of the disease by restoring the body's myelin. Although their oil, named 'Lorenzo's Oil' in honour of their son, is not a cure for the disease, it is now the standard medical treatment for arresting ALD cases.[3]

Few amateurs leave medical science gasping in their wake in this way, for medical research requires expensive laboratories and equipment. But the Odones are far from unique. In April 1993, the defence magazine *Jane's International Defence Review* announced the discovery by a British amateur inventor, Maurice Ward, of a thin plastic coating able to withstand temperatures of 2,700 degrees centigrade – enough to make tanks, ships and aircraft impervious to the heat of nuclear weapons at quite close ranges and hence of overwhelming interest to the military. One is tempted to ask what exactly Britain's defence scientists were doing with the £2,000 million they spend each year on research while Mr Ward – a former hairdresser, dismissed as a crank – was developing one of the century's most important defence weapons in his garden shed?

Many of the people mentioned in this book have conducted fundamental scientific research on their own initiative, without institutional support or funding: people like Nikola Tesla, Wilhelm Reich, George de la Warr, Semyon Kirlian, Harry Oldfield.

Today, the ranks of amateur scientific researchers include investigators of paranormal reports such as poltergeists, psychokinesis, precognition, coincidence, as well as cereologists (those who investigate crop circles) and investigators of unidentified flying objects. Mineral extraction companies are employing 'psychics' and amateur dowsers for metals, oil and water. There has been a massive increase in alternative or complementary medicine in every sphere from the 'new therapies' referred to in Chapter 10 to hypnotherapy, reflexology, acupuncture and faith healing. Some of these alternative therapies are practised by professional medical people; many more are practised by amateurs. Even physics has its amateurs in the form of the 'cranks, housewives and retired professors' who the *Daily Telegraph* feared might be attempting to replicate cold fusion on their kitchen tables.[4]

On one hand such people are often portrayed as eccentrics and derided because they are conducting investigations that have only an amateur status, yet on the other, the very people who criticise them most are those refusing to conduct any professional investigations themselves using the community's scientific resources paid for by the taxpayer. Is it surprising in these circumstances that people are taking matters into their own hands? And is this not a phenomenon that professional science can expect to see a great deal more of in the decades to come? We are already used to hearing the phrase 'alternative medicine'; in future we are going to become used to hearing people talk of 'alternative physics', 'alternative chemistry', 'alternative biology', even 'alternative science'.

Orthodox scientists ought to be deeply disturbed by this failure on their part. Too often they interpret lay skepticism and suspicion not as a failure of their approach to science, but merely as a failure to communicate science's findings, their beautiful understanding, to the public. They busy themselves with Committees on the Public Understanding of Science in the forlorn hope of correcting our incorrigible ignorance with a little judiciously applied public relations. But the truth is that their beautiful understanding is a glamorous illusion, an artifact of reason, and the fact that so many lay people are filling the vacuum of non-activity with their own investigations – and in cases like the Odones and Maurice Ward getting important results – shows the reaction to be merely defensive. Orthodox scientists should recognise this phenomenon for what it is – a symptom of failure of part of their primary function.

Scientists support their claim to be the revealers of a special kind of truth by appealing to the rigorous method they employ: the primary article of faith of their profession, the scientific method. This procedure is the centrepiece of western analytical thought. It is the golden untarnishable truth at the heart of the West's rational philosophy. Centred on the concept of proof, of concrete empirical evidence and repeatability, the scientific method is the closest that human minds have ever approached to eternal truth by rational means.

The scientific method also has its own paradigm – the golden perfect exemplar of its own application and the Royal Road to scientific knowledge. The ideal follower of the scientific method first collects empirical data from the field, being scrupulously objective in taking observations. He records and quantifies those observations, making sure that they are a representative sample of the phenomena under study. Next he formulates a theory to account for those data, recognising Occam's principle that the most economic explanation of the facts is always to be preferred. Finally he designs a crucial experiment – a key test of the theory – in an attempt to prove it wrong. Ideally – according to Sir Karl Popper – this experiment takes the form of an empirical attempt to test a prediction made by the theory which has not yet been confirmed and which, if found to be false, would in turn falsify the theory.

Einstein's general theory of relativity, published in 1915, is often taken as a flawless example of the scientific method in operation. The theory was universally accepted so quickly and decisively because Einstein was able to make several predictions from the theory that were empirically confirmed within only eight years. In 1919 and in 1922, astronomers took advantage of solar eclipses to confirm that light was bent by the Sun twice as much as predicted by Newton's theory. And in 1923, Mount Wilson observatory confirmed that the light from distant stars was shifted to the red end of the spectrum.

Of this latter test for the 'red shift' Einstein wrote, 'If it were proved that this effect does not exist in nature, then the whole theory would have to be abandoned.' It was this commitment, and this willingness to submit to the acid test of empirical proof, that Popper so admired in Einstein. Popper wrote:

> What impressed me most was Einstein's own clear statement that he would regard his theory as untenable if it should fail certain tests.... Here was an attitude utterly different from the dogmatism of Marx, Freud, Adler and even more so that of their followers. Einstein was looking for crucial experiments whose agreement with his predictions would by no means establish his theory; while a disagreement, as he was the first to stress, would show his theory to be untenable. This, I felt, was the true scientific attitude.[5]

In practice, it is not always possible to design such a perfect *experimentum crucis*. Moreover, we have to accept that we are always working with incomplete data. Even more disturbing, it sometimes happens that subsequent data is found that appears to invalidate the original theory, but the data is rejected as 'aberrant' or is attributed to experimental error because the theory has now become so widely accepted.

The most important experiment that led Einstein to formulate relativity theory was that conducted by Albert Michelson and Edward Morley in 1887. The experimenters had constructed an instrument that enabled them to measure the speed of light. They next used their apparatus to see if a ray of light travelling in the same direction as the Earth travels through space faster than a ray travelling at right angles. In fact they found no difference, showing that the Earth has no absolute motion through space and leading Einstein to postulate the constancy of the speed of light, and relativity. However, in the 1920s, one of Michelson's original staff, Dayton Miller, repeated the experiment with much more precise modern equipment giving greatly improved accuracy. Miller announced that this time he *had* found a difference between the two measurements and thus had good evidence that the Earth does have absolute motion through space. Miller's work attracted wide publicity for a while, but because these improved measurements contradicted Einstein, they were quietly forgotten about.[6]

In 1961, *Physics Today* reported a NASA conference on 'Experimental Tests of Theories of Relativity'. The conference heard how the two predictions hailed by Popper were now in considerable doubt: that the red shift of light from distant stars can be derived from a theory more elementary than general relativity, and the same may be true of light bending around the sun. Even the measurements made in the 1919 and 1922 eclipses are now said to be 'equivocal' by some scientists.[7]

The steely self-sacrificing quest for truth that had so impressed Popper suddenly evaporated in the light of these new findings, because no one wanted to question a theory that had proved so useful. Even Popper's perfect example turned out to be flawed on closer inspection. And the quest for truth turned out to be more like a quest for certainty.

Golden though the scientific method may be in principle, there are thus some major difficulties with its application in practice. Chief among them is that science does not usually proceed in the rigorous manner demanded by theory except in school story books or Hollywood film biographies of great scientists. In the real world, the process of scientific discovery is less like a carefully controlled experiment and more like a pantomime of coincidence, accident and adversity. Whatever they may profess outwardly, many scientists construct theories, models, ideas, speculations, that are way ahead of any data they may have, and then set about looking for evidence that might support the theory. When they find a theory that is elegant, beautiful and appears to fit all the known facts, the 'fitness' of the theory becomes more important than the original data. Unsurprisingly, few scientists care to admit to what might be seen as unprofessional conduct in their work. Astronomer and physicist Sir Arthur Eddington is credited with the dictum, 'One should never believe any experiment until it has been confirmed by theory.' Behind the humour, as is often the case in science, there is more than just a grain of truth.

Let me make it very clear that in seeking to identify weaknesses in the application of the scientific method I am not attacking either science or its methods – quite the contrary. I regard the scientific method as the only rational intelligent procedure for attempting to arrive at the truth, just as I welcome the same principles in our system of justice – that when a person is on trial for a serious crime, nothing is taken for granted; that any prejudices we have are put aside; that the accused must be proven guilty by an unbroken chain of empirical evidence beyond any reasonable doubt. What I am concerned about, however, is that just as sometimes an innocent person is wrongly convicted even with all these legal safeguards, so sometimes we reach the wrong scientific conclusions even with the scientific method to help us. That, as Wilhelm Reich observed, 'perfectly exact physics is not so very exact just as holy men are not so very holy'.

This criticism of the scientific method may perhaps seem no more than a pedantic quibble. In reality it is a flaw that goes to the very heart of science. For if the scientific method may be failing us in the manner described above, then what is it that governs the structure of our scientific model of the world? How is it that our knowledge hangs together with such apparent precision, is taught with such confidence, and applied so profitably in technology? The answer to that question is the central proposition of this book: that part of our scientific world model is the creation not of the scientific method in its ideal form; but is a shadow creation which in some respects lives up to the rigorous demands of the ideal of the scientific method but in other important respects does not. Its deficiencies are made up for by the authority of scientists as the arbiters of what is and is not true in the natural world.

If this view is correct, if some part of our accepted scientific world model is a matter of interpretation and rests solely on the authority of some individual scientists – however eminent – despite being on the surface a matter of experimentally confirmed fact, then the important question arises: what part of our apparent scientific knowledge is not really knowledge at all? Is it just a tiny fraction, an insignificant superficial gloss that is added by conjecture? Or is there more to it than that?

From our modern civilised perspective we can look at the belief systems of primitive cultures historically and see that in many cases those who held intellectual authority for shaping the community's beliefs often had little or no warrant for their profession. In the case of many primitive communities, the extent of their scientific leaders' actual knowledge appears very limited by our standards: perhaps a smattering of herbal medicine and a limited amount of astronomical knowledge – enough to predict eclipses of the Sun and Moon.

On this attenuated empirical foundation was erected a fabulous pantheon of animating spirits, demons and other entities, and complex cosmological belief systems to which the whole community was obliged to subscribe. Is our own reductionist belief system any different? And how do we know?

The physics of Empedocles, made popular by Aristotle, would be derided by any conventional scientist today: a physics whose basis is the belief that all things consist of a mixture of earth, air, fire and water. Yet should we honour Aristotle for his scientific insight, and for popularising a scientific system that sustained civilisation for 2,000 years, or should we condemn him for accepting and promulgating superstitious beliefs in the guise of science? And what exactly was it that the Empedoclean paradigm contributed to science? This question is very much the hot potato of the history and philosophy of science, and one that so far few scientists have cared to grasp.

For what if our modern science has as little truth as that of the primitive shaman? Or as little as the science of the early Greeks? Does it not follow that the appearance of power and control conferred by modern science is an illusion? That the intellectual authority of our scientists is little more than a sham? What have we really accomplished other than to be able to kindle a crude atomic fire in the darkness; a fire from which we are compelled to run away like children when sparks fly, as at Sellafield, Three Mile Island or Chernobyl?

If our science contains 80 per cent truth to 20 per cent appearance, then scientists are entitled to their mantle of authority; but if it is 20 per cent truth to 80 per cent appearance, then they are arrogant fools, and we, too, are fools to listen to them and to pay for their vanities.

This gap between empirically determined fact and the totality of the community's scientific beliefs might justifiably be called the myth of science: that some part of what appears to be the substance is actually the shadow. In past centuries, the existence of such a myth may not have been a matter of great public concern. The practice of science itself was largely confined to a small professional community. Many of the great questions were academic in nature – questions as to the structure of the solar system, the size of the universe, the possible origins of life on our planet. Research was largely privately funded and the amount of money spent on research was tiny. In the twentieth century all these conditions have changed dramatically.

As observed at the beginning of this book, the scientific issues have ceased to be academic and have become uncomfortably personal for most of us: issues of global warming, of energy policy, of the environmental effects of industrial and agricultural practices, and of mortal disease on a planetary scale. At the same time, most research is today publicly funded and on a scale of billions a year. Science is no longer the cosy province of the initiated few. Millions who work in industry and commerce possess analytical and intellectual skills and specialised technical knowledge to rival many professional scientists, and are able to evaluate for themselves the effectiveness of our research programmes.

I said earlier that I am not attacking the scientific method, and that is the case. What I am attacking is the presumption of some scientists that we, the community, must accept their conjecture as fact simply because they pay lip-service to the scientific method. Because they are the high priests of science, then anyone guilty of questioning their priestly authority appears to be profaning the scientific method of which they are the guardians. I reject this idea as bogus and demand that all scientists who are supplied with public funds to conduct their research should not merely pay lip-service to the scientific method, but should actually be seen to employ it in their studies and their published results, just as those paid from the public purse to administer justice must also cause it to be seen to be done.

According to Francis Bacon, knowledge is power; but it is more exact to say that knowledge is authority. Power is derived from the subservience of the ignorant to such authority. Let us not grant power to those who lead the community's scientific thought unless they are able to demonstrate that their authority is indeed derived from knowledge.

....

CHAPTER THIRTEEN

The Future that Failed

Humanity has devised many systems of thought to cope
with nature. Yet nature, functional and not mechanical
as it really is, has slipped through its fingers.

WILHELM REICH

To what extent, if any, is science to *blame* for the disasters that have characterised the modern world and differentiated our century from all others? Unsurprisingly many scientists think that their profession is not to blame at all. A representative view is that of distinguished Nobel laureate Sir Peter Medawar who said:

> A former editor of the London *Sunday Express* (a newspaper with a very big circulation) once wrote: 'science gave us the Great War', committing thus the elementary blunder of blaming the weapon for the crime, though not even he, I imagine, would have gone so far as to blame science for nationalism, bungling politicians and ambitious generals such as those who at Passchendale and on the Somme so nearly did in the British Army during the First World War.[1]

Of course, Sir Peter is right in one important respect: the First World War was started by politicians and soldiers, not by scientists. But I think the editor of the *Express* was getting at something much more important with his statement than who pulled the trigger. He was pointing out that the Great War assumed the proportions of a global tragedy because of science. There have been plenty of wars, all started by politicians and soldiers, but most are now trivial and forgotten. In August 1914, the German army was doing exactly the same as it had done forty years earlier when it marched into France. But in 1870, the death toll was measured in thousands and was restricted mainly to professional soldiers. No one speaks of Sedan today in hushed tones or stands at a cenotaph each year in remembrance of the siege of Paris; yet names like the Somme and Passchendale still make us shudder.

The thing that made the Great War 'great', and which burned it into the consciousness of every European nation, which made it arguably the most important historical event this century, and one which changed the face of the whole world for ever, was the slaughter on a scale of millions that science alone made possible in the interval between 1870 and 1914. No amount of nationalism, no political or diplomatic manoeuvring alone could have created such destruction; no ambitious general hungry for glory can do more than give the order to attack. It was science that made it possible for that order to be amplified from personal combat to the deaths of millions, from musket to mass destruction.

The discoveries and inventions that made such a bloodbath possible were in themselves unremarkable, but their cumulative effects were devastating. Rifling of barrels made field guns and small arms more accurate and gave them longer effective range. High-explosive bursting charges that could withstand the shock of being fired from a cannon increased casualties dramatically, while chemical propellants gave further extended range. Maxim's idea of making each cartridge also reload and fire the next round, made sustained automatic small arms fire possible with especially lethal results in trench warfare. All these innovations combined to create unprecedented butchery.

The carnage of the First World War was history's first significant defeat of the human spirit and human values by the irresistible force of machines that make war: the young officers whose breasts were broken by automatic fire could not believe that a lifeless mechanism could conquer human courage and fighting spirit, in much the same way that their colleagues in the science laboratory could not believe that bicycle mechanics could conquer the skies. Science had escaped from the laboratory to become a potent social force. Science was no longer a respecter of persons, no longer a profession for gentlemen.

Scientists, not unnaturally, like to see themselves as honest toilers after truth, the fruits of whose labours are sometimes perverted by unscrupulous politicians. They prefer to believe that weapons are 'morally neutral' and that scientists cannot be held responsible for the use to which their discoveries are put by the unprincipled. While undoubtedly true in many cases, this view is far from being always valid. In 1916, with deadlock on the Western Front, the chief physicist of the Krupp works, Dr von Eberhardt, and the works director, Dr Rausenberger, approached the German war office and explained that it would be technically possible for them to design and construct a gun that would fire a shell some eighty miles – nearly four times further than existing technology – and thus to bombard Paris from the German lines. The military were delighted and commissioned von Eberhardt and his colleagues to design the Paris gun which eventually resulted in considerable loss of life and damage to homes and public buildings.[2] The generals may have killed those people, but the scientist put the gun in their hand and urged them every step of the way.

Wernher von Braun's claim thirty years later that his first V2 rocket 'landed on the wrong planet' was a clever piece of public relations in the service of his claim to be interested only in space travel. But it was he and his team who elected to develop the weapon that killed thousands. Understandably, they wanted to hear the roar of the rocket engine, to see tons of steel rise miles into the sky on a pillar of flame, to feel the exultation of the mastery of such forces.

Von Braun's view was that, 'Science by itself has no moral dimension.'

> The drug which cures when taken in small doses may kill when taken in excess. The knife in the hands of a skilful surgeon may save a life but it will kill when thrust just a few inches deeper. The nuclear energies that produce cheap electrical power when harnessed in a reactor may kill when abruptly released in a bomb. Thus it does not make sense to ask a scientist whether his drug . . . or knife or his nuclear energy is 'good' or 'bad' for mankind.[3]

Twenty-five years after Drs von Eberhardt and Rausenberger approached the German war office with their offer to design the Paris gun, a group of American scientists approached their government and offered to build an atomic bomb. Their fears at what democracy's enemies might do with such a weapon are easily understood. And no one doubts the anguish and sincerity of those scientists like Robert Oppenheimer who agonised over the dropping of the atomic bomb on Hiroshima. But no one, least of all the scientists who worked on the

Manhattan project, could ever have been in the smallest doubt about what they were doing – constructing a weapon of mass destruction to kill thousands of people at once instead of killing them one at a time. Is there really a useful distinction between creating a more convenient means to kill people, and using it to kill them more conveniently?

Not only did science turn war from a professional duel into national carnage, it entirely failed to predict the outcome of its own inventiveness or to issue any warning of what its discoveries meant to the generals and politicians whose surprise and shock at the unexpected holocaust was called bungling with the wisdom of hindsight.

This failure to understand or predict the consequences of its own actions is in many ways the most frightening aspect of science. It is not necessarily the people who built the atomic bomb or the rocket that delivers it to its target that we have to worry about. It is those who have arrogated the role of saviour of their fellow citizens through the exercise of reason and science that we have most to fear.

Take the case of Thomas Midgley, an American chemist and engineer, born in 1889, who believed passionately in the power of science to benefit mankind. Midgley devoted his entire professional career to scientific discovery and to employing those discoveries in the service of his fellows. He first discovered tetraethyl lead, in 1921, and pioneered its use for car engines as an anti-knock ingredient in petrol. Next he devoted his energies and talent to the search for a gas that could be used as a cheap household refrigerant, and he discovered the first of a series of organic gases called CFCs that were marketed under the name of Freon and universally adopted in home refrigerators. Continuing his humanitarian work Midgley pioneered the use of his new gas, Freon, as a propellant in aerosols, where it was especially useful in spreading DDT and other pesticides on a global scale.

Midgley was director of Ethyl-Dow Chemical Company for eleven years and died in 1944, laden with honours from his fellow chemists and the grateful industry and nation that he had served so well. From our perspective today we can see that putting lead in petrol has created dangerously high levels of lead pollution in most of the world's inner cities; that spreading DDT has created super-resistant strains of crop pests while killing off myriad species of harmless wildlife including rare species of bird and insect; and that loosing Freon into the atmosphere has created a hole in the ozone layer that has increased the risk of human skin cancer.

The example of Midgley, and many others referred to in this book, shows that it is impossible in practice for anyone, whether scientist or non-scientist, to have a clear idea what will be the outcome of any particular scientific discovery over even a modest time scale such as a single lifetime. This unpredictability poses scientists with an almost insuperable difficulty. It is impractical – probably impossible – *deliberately* to fail to make discoveries. It is even more impractical to attempt to 'undiscover' them as suggested by the clergyman who wrote to *The Times* in 1945 proposing that the allies voluntarily destroy the 'formula' for the atomic bomb, as though it were a recipe for Christmas cake. Almost everyone would agree that it would be folly to attempt to restrict scientific research in any way when the results could be of such potential benefit to us all – another penicillin or insulin, new synthetic materials or new ways to produce food.

Scientific discovery carries both the promise of great benefits and the burden of great tragedy if misused, but the advantages of the good are generally taken as outweighing the disadvantages of the bad. More precisely, the disadvantages are tolerated because most people believe that there is an additional cumulative benefit that we, the community, derive from a continuing programme of scientific research: that is, a cumulatively greater and greater control over our environment and our lives.

What examples like Thomas Midgley show is that the idea that humanity has learned through science how to control the environment is wholly illusory. Such discoveries merely enable us to *affect* our environment, not to *control* it. Indeed, there is a strong case to be made for saying that the effects of our own scientific endeavours are specifically unpredictable and our world uncontrollable.

It was the post-Newtonian rationalist Enlightenment of the eighteenth century that first fostered the illusion of science as the instrument of control. Those who carried out the French Revolution did so in the name of scientific government, and among the Assembly's first acts on coming to power was the establishment of modern scientific standards of weights and measures, and the promotion of the Académie des Sciences to a pivotal role in the new social order.

Henceforward, believed the revolutionaries, society would be a better place for all men and women because it would be rationally planned, instead of conducted at the arbitrary whim of effete aristocrats, playing at government much as Marie Antoinette had played at being a shepherdess.

The cultural descendants of these scientific Jacobins – in France and every other civilised nation – were still trying to implement their dream of the scientifically planned society two centuries later. In Britain in the 1980s, the Thatcher government believed that monetarist economic policies and the forces of the free market would restore predictability to Britain's economy; while in the United States, the Reagan administration preached the same concept of economic control and planning. A decade later, the whole western economy was plunged into the worst economic recession since the 1930s.

In practice, the revolutionaries were just as arbitrary and irrational in their conduct of national affairs as the aristocrats they replaced; their ability to shepherd their flock as self-deluding as their tragic queen's. Yet the illusion of control through reason has dazzled governments throughout the world ever since.

Up until the Second World War, those who believed in government through the exercise of rational control at least retained a human touch in dealing with their fellow men and women. The introduction of the concepts of science into government was simply an additional dimension in public affairs – it was sold to the voters as better management of the affairs of state, primarily those of the economy. Since the Second World War, however, it seems to have gradually *replaced* humanitarian concepts of government and to have become not a tool to be used *as well as* the principles of parliamentary democracy but a better, surer, quicker method to be used *instead* of democratic methods.

Compare the attitudes of successive governments before and after the war to protecting the community from aggression. In 1939, with Europe on the brink of war, the British government decided that morally it must act to protect the people of Britain against the expected effects of bombardment from the air, whatever the financial cost, and however ineffective. The measures it took were far-reaching and expensive, but they saved tens of thousands of lives. A comprehensive pamphlet advising on air raid precautions was printed and distributed to every home in the country. Air raid wardens were appointed and issued with instructions for coping with the effects of bombing, and stirrup pumps to help deal with incendiary bombs. Gas masks were issued to the entire population including a special design for babies. Later, when war was declared, the government manufactured and distributed hundreds of thousands of steel shelters to homes in target areas and evacuated children to the country. Elaborate preparations were made to put the emergency services on a war footing; preparations that worked well enough when the Luftwaffe attacked to have saved many thousands of lives.

These measures were adopted by the government even though they expected casualties of bombing to be on a nuclear scale – they printed and distributed one million burial forms to local authorities – and even though they expected such measures to be largely useless because there would be less than ten minutes' warning of a bombing attack as early-warning radar was not yet in service.

Some twenty years later, faced with the prospect of a nuclear conflict, the post-war British government made very different preparations to protect the population. Concluding that there was no effective defence against nuclear weapons, and that to attempt to protect millions of people would be both doomed to failure and ruinously expensive the government decided that instead of trying to defend the people, who were doomed anyway if war came, it would instead devote its energies to saving itself. Instead of issuing instructions for safety, which would only be alarmist, the government thus set about spending millions of pounds of taxpayers' money digging secret underground 'Regional Seats of Government' to which it would retire in the event of war.[4]

It is, of course, easy to understand the argument that any attempt to protect the population against nuclear attack would merely be Quixotic; that there is no defence against thermonuclear weapons; that there would be negligible warning of atomic destruction; and that the situation was thus entirely different from 1939. In fact, these very arguments were used *in* 1939 when it was widely believed that – in the words of former prime minister Stanley Baldwin – the bomber would always get through. This very argument was put forward by the rationalist faction of those against spending money on expensive pamphlets and gas masks and stirrup pumps but who, mercifully, were over-ruled by those who believed that the primary function of government was to defend the community.

The plain fact is that something deeply unpleasant happened in British public life in the twenty years between 1939 and 1959 that signalled the end of a sense of duty and responsibility by the peacetime government towards the people, and its replacement by a rationally justified new mood of selfishness. It is a tendency that persists in government to this day. It is a profound cynicism and selfishness that permeates every level of society, but it starts at the top. And its name is scientific government. It is the culmination (some would say the inevitable outcome) of the rational political sentiments that swept Europe and America in the eighteenth century: reason alone can improve the lot of the community, therefore reason is supreme – more important even than the people it is intended to serve, considered as individuals, since life is ephemeral while reason is eternal.

In this brave new rationalist world, the function of government is not to protect the members of the community it serves. It is something subtly different: to promote the greatest good for the greatest number. When reason dictates that the greatest number whose good can be achieved is merely the handful of government officials who can be accommodated in the nuclear fall-out shelters, then the government is still seen to be obeying the eternal principles of reason, in the name of humanity.

This something nasty that has crept into public life in the past fifty years has tainted the whole liberal-humanistic western intellectual tradition, like a dead rat under the floorboards. It has preserved an outward appearance of benevolent tolerance in the sciences and in academic life generally but beneath the surface it has hardened into a cold authoritarian cynicism that springs to spit ridicule at anyone with the audacity to question any of its beliefs.

This influence has spread so far that neither academic nor scientific dissent is easily tolerated. When Professor Sheldrake proposed an innovative biological theory, the editor of the world's leading scientific journal was able to call for the book to be burnt, without causing even a raised eyebrow. When Doctors Fleischmann and Pons had the effrontery to discover a new source of energy, they were publicly ridiculed and humiliated.

Why on earth should such a change have come about? What historical forces can have caused it? Perhaps the most striking contrast between science today and science in the eighteenth and nineteenth centuries is the loss of the spirit of optimism. In the centuries after Newton, Pascal, Franklin and Goethe, public affairs were infused with a political spirit that was both democratic and scientific in a way that people believed heralded a bright new age. The marriage of politics and science would transform society so that the old ills of disease and poverty would simply wither away. Even the dark side of this dream held out some hope: in Shelley's novel, Frankenstein's monster is a frightening innovation but an innovation that can be of service to humanity provided we master the ethical questions involved. A more responsible Baron Frankenstein could be a public benefactor rather than a public menace.

The most important part of this socio-scientific thinking is once again the idea of control: that knowledge is power, and scientific knowledge confers the power to control the world. The ability to employ scientific discovery in a directed way meant that after millennia of being at the mercy of a fate as capricious and as trivial as a smallpox virus or a hungry locust, human society now possessed the power to shape its own destiny, its own future. This idea was still going strong when Thomas Midgley put lead in petrol, Freon into refrigerators and DDT into aerosols. By the time the world woke up to Rachel Carson's *Silent Spring* in the 1960s, the dream was dead.[5]

Like many dreams, the idea of science as the saviour of mankind did not end quickly and all at once. Its dissolution has been a gradual process, partly because science itself always seemed to be able to pull another rabbit out of the hat, leaving the audience gasping in amazement: television; nuclear energy; semiconductors; space travel; laser surgery; computers; genetic engineering. All seemed to be pointing in the same direction, that of greater control: control over our lives, and over our environment.

But the more 'controlled' the environment became, the more it seemed to be polluted, and the worse the ecological disasters. And the greater the 'control' over our lives the more frightening the diseases, the wars, the crime, the economic depressions.

The dazzling discoveries of science came as a series of waves or cycles – the electromagnetic, the nuclear, the microbiological, the information revolution – as each succeeded the other it was eagerly taken up by scientists and other rationalists as the primary instrument of control. With each new wave, scientists believed that they had finally placed their hands on the ultimate levers of power. But each time, just as they thought they were finally in the driving seat, nature seemed to reassert her wild, unpredictable side and scientists were rebuffed again and again; compelled to turn in disappointment and frustration to the next 'wave of the future'. One after another the new technologies, new discoveries and inventions they spawned first promised to yield up the innermost secrets of nature, and then dashed science's hopes, leaving the community still powerless to predict or control future events; every bit as much in the grip of impersonal natural forces as the people of medieval times.

At the end of the nineteenth century, those who believed that electricity had the power to move and shape the entire world can never in their wildest dreams have conceived anything so fantastic as the ultimate legacy of Faraday and Maxwell – the 'flame effect' electric fire, the pop-up toaster or the electric toothbrush. Those visionaries who in the 1940s and 1950s perceived in nuclear power mankind's salvation failed to imagine the radioactive white elephants of the 1990s, derelict nuclear power stations with their decommissioning costs of billions paid for by a society with 3 million unemployed and thousands sleeping on the streets in cardboard boxes.

The last gasp of the science-as-control movement came with the information revolution in the 1960s. In the digital computer, science believed that it at last had the instrument to *model* the world and hence *predict* the future course of events. Moreover, the power of electronic information processing is so awesome that it would enable us to analyse and model even the most complex systems – the economies of nations and even the structure and processes of living organisms.

If knowledge is power, and the raw material of knowledge is information, then the road to scientific control must lie in acquiring and processing vast quantities of data. After all, if we know *everything* there is to know about a thing in the most minute detail then we can predict with perfect accuracy how it will behave in future.

In the early 1970s, the largest and most powerful computers were those employed in modelling such complex systems. The biggest IBM machine in Britain for example (or the biggest that could be publicly talked about) was the machine installed by the Meteorological Office at Bracknell to model the world's weather system and hence predict weather more accurately. Perhaps here was even the first tentative step towards that age-old dream of weather control?

At the treasuries in London, Washington, Paris and Bonn, equally complex machines were fed with mountains of statistics on production, imports, exports, interest rates and a thousand other details. Economists and civil servants stood by expectantly, anxious to learn what the future held for their nations' economies and what tiny adjustments would be necessary to manipulate those economies in the desired direction of industrial growth, full employment, low inflation and a positive international trade balance.

Sadly, the results of these attempts to grasp nature through its minutiae are only too familiar. We continue to live in a world that is at once banal in its sameness and terrifying in its chaotic unpredictability. Hurricanes, droughts, snowstorms still arrive with their accustomed unexpectedness, while the world's economies continue to defy the best efforts of economists.

The way in which institutional science, funded with public money, has sold this perpetual failure to the public that pays for the research is as essential steps to learning: a necessary struggle over the obstacle course that nature erects in our path. These failures may be costly, they may be disappointing, but they are unavoidable teething troubles if we are ultimately to get our hands on the levers of power, a long-term goal for the attainment of which almost any sacrifice is considered worthwhile. In reality, the failures represent an addiction to larger and larger injections of research funds.

This permanent and recurring failure to learn, or unwillingness to learn from our mistakes, characterises institutional science in the twentieth century. The scientist who believes in the power of science to control is like the compulsive gambler who claims to have learned his lesson and henceforward will be a model citizen – now that he has mastered his gambling urges. In reality, no amount of experience of the unfathomable chaos of nature can disconfirm his belief in rational control. He is not so much like the reformed addict as like Mr Toad, who will promise anything to his friends, so long as he can once more get behind the wheel of a car and take to the open road. Just one more billion pounds in research funds, and the secrets of the universe will be ours – we will have a Theory Of Everything.

In his *Critique of Pure Reason*, Kant observed that:

> It is the land of truth (an attractive word) surrounded by a wide and stormy ocean, the region of illusion, where many a fog-bank, many an iceberg, seems to the mariner on his voyage of discovery, a new country, and while constantly deluding him with vain hopes, engages him in dangerous adventures, from which he never can desist, and which he never can bring to a termination.[6]

The island that Kant saw as so treacherously guarded is the land of pure understanding.

Where will science turn next? If electricity, chemistry, space research, nuclear energy, computers, have all failed as keys to the future, what is left for scientists to turn to? The answer to many scientists appears to be microbiology and, specifically, genetic engineering. Returning conceptually to the roots of reason in the late eighteenth and early nineteenth centuries, modern science has rediscovered the spirit of Baron Frankenstein.

By splicing bits of DNA into the embryonic cells of various plants and animals, microbiologists have succeeded in altering their physical characteristics to suit the consumer. Thanks to genetic engineering we now have domestic livestock and vegetables that the Baron would have relished at his table: pigs with an extra nipple for suckling more young or bred with organs ready for transplanting to humans; rainbow trout with human genes that cause them to grow bigger and more rapidly; tomatoes with the genes from a flatfish that can withstand frost.

Extraordinary feats of this kind regularly make headline news and promote the impression that microbiologists have finally broken through the mystical barrier that surrounds living organisms and have seized control of the levers of life. Genetic engineering (which is perfectly real) tends to be publicly perceived as part and parcel of a vast microbiological project whose aim can be summed up as the replication – and eventually creation – of life in a test-tube.

The idea has been best expressed by Frank N' Furter in Richard O'Brien's *Rocky Horror Show*. Frank tells his unexpected guests, Brad and Janet, that they 'are to witness a new breakthrough in biochemical research – and tonight paradise is to be mine'. Explaining how an accident led him to make the research breakthrough that had eluded him for years, Frank tells them: 'That is how I discovered the secret. That elusive ingredient, that spark that is the breath of life. Yes, I have that knowledge. I hold the key to life itself!'

While this is parody, the sentiments come uncomfortably close to echoing what most of us have been led to believe of microbiology, since Watson and Crick deduced the structure of the DNA molecule and discovered the basis of the genetic code in 1953.

In promoting microbiology as the science of the future and the ultimate tool of control, science has popularised a cluster of related ideas: that the processes of microbiology are at root simple chemical processes that can easily be replicated in the laboratory; that this simplicity provides a natural explanation of how life came into being spontaneously and of how we ourselves can, in principle, create and modify life at will; that the genetic engineering that puts fish genes into tomatoes is but the first simple step on a ladder of progress that will ultimately enable us to create anything in the laboratory that our imaginations can picture and our system of ethics will permit. The rhetoric is that of Frank N' Furter.

These ideas are widely promoted by microbiologists and their popularisers to the extent that most non-scientists (and even some scientists) today accept them. Yet all are demonstrably false, and self-deluding. The myth that science is able to replicate natural biological processes in the laboratory was propagated (doubtless in good faith) at an early stage by some of microbiology's most eminent researchers. Writing in his 1966 book *Of Molecules and Men*, in which he set out to show how simple and mechanistic biological processes are, Francis Crick questioned whether there could be any physiological seat for a 'vital principle' governing microbiological processes. 'It can hardly be the action of the enzymes,' he says, 'because we can easily make this happen in a test-tube. Moreover most enzymes act on rather simple organic molecules which we can easily synthesise.'

Crick adds that: 'It is true that at the moment nobody has synthesised an actual enzyme chemically, but we can see no difficulty in doing this in principle, and in fact I would predict quite confidently that it will be done within the next five or ten years.'

Going on to discuss mitochondria (bodies in the cell that contain DNA), Crick said:

> It may be some time before we could easily synthesise such an object, but eventually we feel that there should be no gross difficulty in putting a mitochondrion together from its component parts. This reservation aside, it looks as if any system of enzymes could be made to act without invoking any special principles, or without involving material that we could not synthesise in the laboratory.[7]

Similar words are still being used a quarter of a century later by microbiologists. The very idea that microbiological processes are at root simple well-understood chemical reactions is a key tenet of modern reductionism. Yet the fact is that not one significant organic molecule, let alone cell component, has yet been synthesised from scratch. Despite all the great achievements of those years, the early hopes that important elements of organic life could be synthesised relatively easily or in the near future have evaporated. No one has synthesised from non-living materials anything remotely as complex as an enzyme, or indeed any kind of protein molecule, nor any part of such a molecule, let alone a complex body like a mitochondrion. *All* genetic experiments, no matter how sophisticated they appear to be, involve merely tinkering with already living things and are no more advanced in principle than cutting the leg off a star fish or salamander and watching another grow in its place. How this happens remains a complete mystery to microbiologists.

It is true that through a variety of ingenious methods, principally those of X-ray crystallography, microbiologists have completed the immensely complex and difficult task of deducing the three-dimensional structure of a few protein molecules. But there is a world of difference between describing the structure of such a molecule and being able to make one in a test-tube. This act of creation demands not merely organic chemistry but something else as well.

As far as most people are concerned it is the 'something else' that is the interesting part: the important part; the part we are curious to know about and to have explained; the part that is the actual subject of microbiology; what it is that makes a living organism out of a handful of dust. Yet many scientists are content to assert, without any empirical basis to support them, that the mechanisms of life have been discovered, that – in the words of Frank N' Furter – we hold the key to life itself.

Anyone, whether scientist or non-scientist, who examines the record of the past hundred years dispassionately will find that, despite the enormous strides made by science, and despite its incalculably great contribution to human development, most of the claims made for science by those who insist it is the instrument of control are at best unsubstantiated, at worst entirely bogus.

The truth is that nature is infinitely more subtle and complex than our current scientific model admits, while our scientific methods of controlling nature are by comparison primitive and, in the long term, usually ineffective. In the words of Wilhelm Reich, nature has somehow slipped through our fingers. Scientific methods had apparently largely eliminated diseases such as tuberculosis, but in the 1990s more dangerous highly resistant strains are again epidemic in every country around the world including the most civilised nations. In April 1993, the World Health Organisation declared the new resistant strain of tuberculosis a global emergency that may claim 30 million lives in the next decade. The use of chemical pesticides was thought to have rid the world of dangerous crop pests. Yet the modern world is now being plagued by new generations of resistant insects that not only spread disease but also destroy crops more voraciously than before mankind intervened. DDT, for example, is now known to be carcinogenic in humans and of the thirty-six species of insect that are cotton pests, twenty-five are now immune to DDT.

Similar contradictions between the dreams of reason and the realities of rational intervention are equally evident in the social policies of the 1990s. In previous generations only George Orwell could have imagined a 'Ministry of the Environment' that is *actively* planning to destroy the last example of wild-wood in southern England at Oxleas Woods, and one of Hampshire's most beautiful and historically important sites at Twyford Down by driving motorways through them. The irrevocable loss of quality of life, the human cost, of these measures is of secondary importance because reason ordains that the roads are necessary to effect the greatest good of the greatest number.

Anyone who sits down to cheer themselves up by counting their blessings will find that an unendingly long list of those blessings are the products of science and scientists: remedies for many of the worst diseases; machines that have eliminated the life-shortening drudgery of existence and that have provided a mechanism for the creation and distribution of wealth on a historically unprecedented scale. Yet even while counting their blessings, no one can ignore the shadow creation of scientism: the authoritarian and anti-human application of reason to problems that are essentially not amenable to rational analysis or solution.

As well as providing the bounteous gifts described above, it was science in its shadow form that put the Birmingham Six in jail for sixteen years; that poisoned the inhabitants of Bhopal; that covered Sellafield, Three Mile Island and Chernobyl with radioactive dust. And it was scientific rationalism that caused the British government to protect only itself against the effects of nuclear war, to turn mentally ill people out to sleep on the streets; and to drive a motorway through the last wild-wood in southern England in the name of a rational public policy.

These are not isolated abuses, or the acts of a few misguided individuals. They are the inevitable outcome of elevating science to a position of authority as the repository of important eternal truths, and to a position as the fount of wisdom for regulating and governing important practical aspects of our society. Vital though science is in our society, and great though its contribution may be, it merits nor warrants either role.

CHAPTER FOURTEEN

A Methodical Madness

*My own suspicion is that the universe is not only
queerer than we suppose, but queerer than we can suppose.*
J.B.S. HALDANE

In Shakespeare's *Hamlet*, when Polonius wishes to humour the prince,
he willingly agrees to whatever half-baked observations Hamlet cares
to make.

'Do you see yonder cloud that's almost in shape of a camel?' The
prince asks Polonius. When the old man concurs, Hamlet changes his
mind: 'Methinks it is like a weasel.' Polonius agrees that it is 'backed
like a weasel.' 'Or like a whale?' ventures Hamlet. 'Very like a whale,'
Polonius readily agrees.

To Hamlet this willingness to agree to any interpretation of the
cloud's shape was evidence merely of the courtier's belief in his feigned
madness. 'They fool me to the top of my bent,' he confides in an aside
to us, his audience.

This exchange is fascinating because it embodies the very essence of
the difference between orthodox reductionist science and those who
see the world in less black and white terms.

As children we all lay on the grass and played Hamlet's game
'seeing' camels, weasels, whales and a hundred other shapes as the
clouds drifted across summer skies, comparing notes and trying to
'see' what our friends could see. Hamlet, the rational avenger of his
father's murder, takes it for granted that the shapes are unreal,
existing 'only' in the imagination, while Hamlet the pretend madman
– with 'method in his madness' – seems to think the essence of insanity
is impressing his own perceptions on the outside world. The question
is, who is right? Is the cloud 'really' shaped like a weasel or is the

connection only in our minds? What does it mean to say that the shape exists only in our minds?

Reductionists think they have a simple explanation for this phenomenon. The brain is a computer, loaded with pattern recognition software that merely sorts through memory until it finds a match or close match to the object projected on to the optic nerve by the eye. The 'program' learns to recognise patterns by successive comparisons. In fact, we know from experiments that the brain does *not* work in this way, but reductionists remain optimistic that some variant of this relatively simple process will eventually turn out to be true: that memory has a physical substrate and that the brain, although it doesn't work like the digital computer on my desk, is some kind of information processing mechanism, and that the recognition of patterns is no more than a matter of comparison.

The problem with this view is demonstrated by the examples in earlier chapters. What we perceive is not necessarily determined by the signals occurring in the optic nerve, but by what we expect to see – signals occurring within our brain. A black four of hearts will usually be seen as a red card, or as a four of spades. So how does the hypothetical pattern recognition software know what to search for? And how does it know when it finds it? There appears to be some higher function, a global function that directs brain processes from a 'mission control' level. This function seems to be involved in some of the strangest and least intelligible of human abilities: it 'feels uncomfortable' when shown anomalous playing cards; it 'knows' how to recognise weasels or whales independently of the stored patterns of their shapes in memory; it may even acquire knowledge by extra-sensory means – but how?

One scientist who came close to grasping the answer to this question experimentally was Princeton University's Wilder Penfield. In his 1975 book, *The Mystery of the Mind: A Critical Study of Consciousness and the Human Brain*, Penfield described an experiment whose consequences astounded him, and which might have made reductionists everywhere pause for thought, but which remains ignored.

Penfield described an operation carried out on a patient he called C.H., who suffered from epilepsy due to the common cause of lesions in the temporal lobe of the brain. But the lesions in this patient's brain were dangerously close to where his major speech centre should be: any damage to this region could cause permanent impairment of the patient's speech ability. In order to avoid any such accident, Penfield decided to map out the exact geography of the patient's speech area, and to do this, he probed the area with a sensitive electrode that used a weak electric current to inhibit speech by temporarily inducing 'word blindness'. This procedure may sound a little alarming but the brain is not actually sensitive to touch, so although the low level of electric current inhibited his speech, the patient felt nothing and was unaware of the inhibition until he was shown a picture of an everyday object and asked to name it. By repeatedly probing the speech area and asking C.H. to name pictures to see if his speech was inhibited or not, a complete map of the area could be built up.

It was what happened next that astonished Penfield. He applied the electrode to C.H.'s speech area and showed him a picture of a butterfly. The patient snapped his fingers apparently in exasperation and seemed to be searching his memory. When the electrode was removed, C.H. immediately said, 'Now I can talk. Butterfly. I couldn't get the word "butterfly" so I tried to get the word "moth"!'

Although the patient's verbal mechanism was electrically inhibited, and he was temporarily 'word blind', the patient could nevertheless perceive the meaning of the picture of the butterfly and could make a conscious effort to recall the correct name. Finding he was unable to recall it, he turned back again to his interpretive mechanism – which was clearly not being affected by the electrode and so was not located in the speech centre – and searched for a second concept that he felt was the closest thing to a butterfly.

'The patient's simple statement startled me,' Penfield wrote. 'He was calling on two brain-mechanisms alternately and at will . . . I can only say the decision came from his mind.' Penfield came to believe that some aspects of brain function such as memory may be governed by something higher than the brain itself.[1]

This very idea is strictly taboo to reductionist scientists, because it smacks of mysticism, religion and barmy cult beliefs. What can be higher than the brain? And where could it possibly reside? We seem here to be straying dangerously close to ideas like 'soul' or 'spirit', trying to import magic into biology.

The attraction of scientific reductionism and the motive for wishing to banish such metaphysical notions is not difficult to understand. Science has provided reasonable naturalistic explanations for many of the most important philosophical questions: How did life arise? What is mankind's position in the scheme of things? What holds together the physical fabric of the world and keeps the stars in their courses? These answers seem final, or close to final, after thousands of years of doubt. But it is not the finality of these explanations, or the quality of evidence that supports them that makes them so acceptable. The key word in this explanation of the causes of scientific reductionism is 'reasonable'.

To the scientists of the Babylonian civilisation, it seemed reasonable to believe that the Earth was flat and was held up by elephants standing on a giant sea turtle – even though their astronomy was highly developed and they had observed the curvature of the Earth's shadow moving across the Moon during eclipses. They held this view because they could not imagine a plausible alternative theory. The idea of a flat Earth held up by elephants was the *most reasonable* explanation available. Flatness seemed to fit their everyday experience, and, although highly improbable, elephants were far less improbable than any conceivable alternative. Yet, because it was based on faulty evidence, it was actually only a superstitious belief. What appeared the most reasonable view was really completely unreasonable.

The flat Earth theory was rejected by Greek scientists such as Pythagoras, Hipparchus and Aristotle who observed that the Sun and Moon were spherical and reasoned that the Earth must be too. Once the flat Earth viewpoint was deprived of the appearance of being reasonable, its wildly improbable nature became obvious, and it seems amazing to us today that anyone could have believed in such a theory, however limited their scientific knowledge.

I believe that something very similar is true of parts of western science today. It actually contains some wildly improbable theories – as improbable as elephants holding up the Earth. Yet these theories appear to represent a reasonable view because they offer a natural-sounding mechanistic explanation that seems consonant with common sense and our essentially limited experience and understanding of the world.

Whole areas of the western scientific model come in this category: theories that seem as solid as rock and, indeed, are the foundations of much of western thinking, yet in reality are at best unsubstantiated and at worst no more than superstitions: Darwinism,[2] Freudianism,[3] Marxist historical and economic analysis are prime examples of flat Earth beliefs that have been exported the world over. But why should any rational person – let alone a trained scientist – accept such beliefs?

One especially strange aspect of belief in western culture is that we habitually use the word belief to mean two entirely different things depending on whether we are speaking of belief in an everyday sense (I believe in parliamentary democracy) or in the scientific sense (I believe in the atomic theory of matter). It is normal in our culture to take the second statement as meaning that the empirical evidence and theoretical background of atomic theory are such that any rational person who analyses the facts must be compelled to accept the theory. We also think that this process of 'scientific' acceptance is different in kind from the ordinary acceptance of everyday things: a person might be right or wrong to believe in the value and effectiveness of parliamentary democracy because it is a matter of opinion, but he or she cannot be wrong to believe in atomic theory because it is a matter of fact.

Yet the *psychological* process of acceptance is actually the same in each case: it rests simply on the fact that the conclusion seems to be irresistible, even to the well-informed and educated mind. This appearance of being irresistible can in itself be a self-evident justification for belief – just as it is 'obvious' that two and two must make four, and just as it was obvious to Babylonian scientists that the earth is flat.

The problem that this psychological process can present, as we saw earlier, arises because our perception – and hence what appears obvious – is to some extent determined by our beliefs. It means that all observers, scientists as well as savages, employ a kind of mental inertial guidance navigation system which takes over our routine mental processing; an intellectual autopilot whose perpetual heading is the star of our convictions, and which filters our perceptions to ensure that they conform to those convictions. This mechanism has been graphically described by Arthur Koestler in his book *The Ghost in the Machine*.

> I can speak . . . with first-hand experience, based on seven years of membership in the Communist Party during Stalin's terror regime. In writing about that period, I have described the operations of the deluded mind in terms of elaborate manoeuvrings to defend the citadel of faith against the hostile incursions of doubt.

Travelling through Russia, Koestler writes that he saw the results of forced collectivisation of the land.

> I reacted to the brutal impact of reality on illusion in a manner typical of a true believer. I was surprised and bewildered – but the elastic shock absorbers of my party training began to operate at once. I had eyes to see and a mind conditioned to explain away what they saw. This 'inner censor' is more reliable and effective than any official censorship . . . it helped me overcome my doubts and to re-arrange my impressions in the desired pattern. I learnt to classify automatically everything that shocked me as the 'heritage of the past' and everything that I liked as the 'seeds of the future'. By setting up this automatic sorting machine in his mind, it was still possible in 1933 for a European to live in Russia and yet remain a communist.[4]

This, of course, is the phenomenon in an extreme form and most of us have a tendency on reading such words to think: I could never be so blind. Yet this is exactly how all our beliefs are constructed, and scientific beliefs are formed psychologically in a way that differs not at all from the most extreme ideological communism or the most intemperate religious fundamentalism. It is not, finally, the facts that matter; it is the pattern those facts make in our minds. The important question here is: what exactly is it that determines the patterns that accommodate the facts?

It is as though our perceptions reach our minds through a screen – a matrix that is dynamically adaptive to our world view and that can selectively modify the contents of our field of vision in the service of that world view. That this is literally true in some circumstances was shown by the research of Bruner and Postman, and that of Festinger described in Chapter 8.[5,6]

Take, for example, the night in August 1877 when the director of the United States Naval Observatory, Asaph Hall, turned his telescope in the direction of the planet Mars. No one knows what Hall's thoughts and feelings were as he began his observations, but a few minutes later he was confronted by 'fear and terror'.

Hall had been tempted to turn his telescope in this direction by centuries-old speculations that Mars, like the Earth, might possess a moon. The idea of a Martian moon had cropped up repeatedly in both factual astronomical works and in fiction as far back as Galileo's day. In his satirical 'Journey to Laputa', published in 1726, Jonathan Swift had even described the distances and periods of rotation of not one but two Martian moons – a joke at astronomy's expense which was to prove astonishingly accurate in every detail.

Stimulated by such conjectures, the world's astronomers had trained their largest telescopes on the mysterious planet many times but found no trace of a satellite. But on the night in 1877 that Hall looked for them, they suddenly appeared. Indeed, their appearance was so sudden and unexpected after years of fruitless observation that until they were photographed close up by the *Mariner 9* space probe in 1971, some astronomers had seriously begun to suspect they might be artificial satellites launched by Martians on the night that Hall first observed them. What Hall himself felt about his discovery is perhaps signified by the names he bestowed on the two moons: Phobos and Deimos – Fear and Terror – the sons of Mars.

Although he never publicly admitted as much, I suspect that Hall felt in one brief, irrational moment that he had invented the moons. That he had wanted them to be where Swift had said they would be and his wish had been magically realised. Many people have had this irrational idea creep unexpected and uninvited into their minds at some time or other. Those people whose business is discovery – physicists, astronomers, biologists, explorers, archaeologists for example – are especially vulnerable to the feeling of inventing the things that they discover, particularly when they find precisely what they are looking for. Such people often speak of having an instinct or a sixth sense for discovery.

This almost mystical experience strikes different people in different ways depending, I suppose, on how far they are prepared to question the nature of their own beliefs. Those scientists who believe that reason alone can explain the world are likely to dismiss such experiences as unscientific, irrational, and no more than a temporary lapse into the unschooled thinking of childhood, the primitive associative thinking that Jean Piaget called syncretic.[7] Some scientists – such as Albert Einstein and Sir Oliver Lodge – have speculated that rational science reveals only the external appearances of some deeper reality. A few even bolder thinkers have recognised a non-rational dimension of the physical world and of human affairs and have attempted to come to terms scientifically with this mysterious deeper reality – men like physicist Wolfgang Pauli and psychologist Carl Jung who jointly proposed a theory of synchronicity or an acausal connecting principle to account for phenomena which do not yield to classical science.

Jung, for instance, was struck by coincidences which exceed all rational explanation. He often told the story of how, when a woman patient was describing to him her dream of a giant golden beetle, they were interrupted by an insistent tapping at the window. When he opened the window, a large golden scarab beetle flew into the room. Jung's response to such phenomena was to say that some experiences are not merely irrational, they are beyond reason and may only be grasped intuitively. His theory of synchronicity proposes that nature is connected up in some mysterious non-rational manner inaccessible to our rational minds, but accessible to our unconscious.

Despite the stature and achievements of men like Jung and Pauli, very few scientists would dare to admit anything remotely resembling the theory of synchronicity into their everyday calculations, not only from fear of ridicule by their scientific colleagues but also because such an admission threatens – or seems to threaten – the very basis of a rationalist science. Yet, while science dares not contemplate any theory as anarchic as synchronicity, it cannot entirely suppress an uneasy sense of something big and mysterious lurking on the threshold of consciousness: physicist Otto Hahn writing 'It may be that a strange series of accidents renders our results deceptive', and Haldane voicing his suspicion that the universe may be 'queerer than we can imagine'.

We live in a topsy-turvy world where nothing is what it seems, thanks to an exclusively rationalist science. I cannot help feeling that, like Hahn and Haldane, astronomer Hall was unconsciously trying to tell us something of the sort with his discovery – perhaps that from a Martian viewpoint (that is, a non-rational or non-human viewpoint) lunacy is the business of extending human knowledge through the creative exercise of reason on our already rationalist world view. As long as all the 'facts of nature' can be made to hang together by reason, then science succeeds. When they begin to fall apart – and they are beginning to fall apart in our day – then science must resort to the lunatic's refuge of denials and rationalisations.

The scientific view of the world has been an exclusively rationalist view for only a relatively short time – probably less than two centuries. In Newton's day, and even in James Watt's time, reason was merely one tool for exploration and explanation; the world remained fundamentally mysterious and capable of inspiring awe in scientist and non-scientist alike.

Anything seemed possible, discovery unlimited, at the time when Faraday wrote that 'Nothing is too wonderful to be true, if it be consistent with the laws of nature.' But because of the enormous success of technology in the nineteenth century, reason has today been granted a monopoly on the right to illuminate and inform our view of the world. Rationalist science has been immensely effective. But it has preferred to ignore the growing clamour of the inexplicable phenomena that loiter at its door like a band of ragged orphans begging for admittance. No matter how accomplished rational science has become at explanation, there are always further outbreaks of table rapping, spoon bending, and other outrages against science, almost as though Discordia or Eris, the goddess of chaos, remains one jump ahead of scientists who try to deny she reigns supreme.

But how can science be expected to get a global perspective on phenomena which at present are incomprehensible in practically all respects? Let me turn the question on its head. Why – exactly why – are we having so much difficulty in coming to terms with and explaining anomalous phenomena? After all, we have large amounts of data from a number of fields of research, and no shortage of subjects to study. Yet when any serious debate is begun respecting this data, any attempt to draw conclusions from it – even rational attempts – are resisted almost fanatically.

Take just one example. The statistical studies of planetary positions at time of birth carried out by Michel and Françoise Gauquelin, mentioned in Chapter 10, provide empirical evidence in support of some of the traditional claims of astrologers (the birth charts of sporting champions are correlated with certain positions of Mars, for example). The orthodox scientific response to such claims is usually to demand more and more statistical studies. In fact the Gauquelins' work has so far been replicated seven times, so let us for the moment accept that a *prima facie* case has been established by existing studies. The question then becomes: if planetary positions are correlated with some aspects of human affairs in some predictable way, why do we find this discovery so difficult to accept? And why is it that science has failed to notice and make use of this relationship all these years? After all, the effect is no more surprising or offensive to our scientific world view than the idea that bodies attract each other at a distance in an invisible, intangible way; and it is considerably less difficult to swallow than some of the effects predicted by relativity theory, such as the slowing down of time with increasing speed. So just what is it about astrology that offends us to such an extent that we should reject the idea so forcefully?

To say that astrology is unacceptable because it is unreasonable or irrational will not do. None of the facts of science – even though they have been arrived at empirically or inductively – can claim to be rational in themselves. If it is observed in the particle accelerator (as it has been) that an electron grows in mass as it accelerates to speeds approaching that of light, then that fact by itself is neither reasonable nor unreasonable, neither rational nor irrational, it is simply so. That the effect is predicted by scientific theory does not make it a rational fact – it merely confirms that scientists are thinking along productive lines. The processes by which people design cyclotrons, observe particles and think about what they see may be rational or they may not. But the facts of nature themselves cannot be described by purely human attributes.

This is often far from clear. Not unnaturally, scientists like to think that the predictions of their theories are rational expectations. But only nature is the arbiter of what is and what is not. Before the first Moon landing in 1969, it was quite possible for someone to entertain legitimately the hypothesis that the Moon was made of green cheese. No one could be contradicted on scientific grounds for holding such a theory, even though it seems highly improbable, for only the examination of rock samples can settle such a question. The important point here is that, by the same token, someone who – before 1969 – wished to entertain the hypothesis that the Moon was made of rocks like the Earth's, held a view which was neither more nor less rational or scientific than one who held the green-cheese theory. Yet this theory appeals to us naturally as sounding more 'reasonable'.

Nor is this all, for if NASA were to send spacecraft to a million moons circling a million remote planets, those observations would not affect the probability that the next moon we examine will be made of cheese or any other substance. David Hume pointed out in 1748 that an empirical observation that something is so, does not logically entail that the subject will continue to be so when we next observe it. Science is unable to say that water does not burn – merely that it never has, so far.

Hume's discovery that the whole of science is vulnerable to the fallacy of induction has had far-reaching effects. Hume has been accused of having given up science to become an irrational mystic. But this is to misunderstand the real nature of Hume's discovery. What he showed was simply that the facts of nature cannot be considered to be rational or irrational. People's minds and actions may be rational, but an apple falling from an apple tree cannot. I believe that science has felt itself reeling from an imaginary blow ever since Hume, simply because many scientists wish to insist on believing that the universe itself is rational. Hume pointed out to them that their belief is incorrect, but this should not be discouraging to one who is truly rational, since reason is simply our peculiar way of grasping the universe and exploring its make-up – the map and not the territory.

Instead of disowning Hume as a mystic, science should acclaim him as the man who set our minds free to employ reason without the crippling restriction of a primitive superstitious belief that the universe is in some magical way a cosmic analogue of our own minds. It should not matter to a truly rational science whether the Moon is made of green cheese or rock. It should be enough that one or another is true. And if it turns out that both are true, then that, too, would be a fact which science must joyfully accept as expanding the horizons of human knowledge, however inconvenient it may be.

I believe that it is this superstitious belief in a rational universe which is the unconscious motivation for rejecting anomalous natural phenomena. Scientific anarchy seems to be looming in the wings, waiting impatiently to make an entrance just as soon as we admit the truth of our observations; the reality of our experience. Admit that being born 'under' Mars increases a child's chances of being sporting – runs the unspoken fear – and science will fall to the ground amidst the ruins of a once-rational cosmos. This fear is entirely groundless. However marvellous or bizarre any particular fact of nature should turn out to be, the cosmos remains what it always has been. It cannot cease to be rational, because it has never been rational. It simply is.

The method of science is to look for pattern, for order in the cosmos. Amongst the welter of information reaching our senses, we look for that which stands out by reason of its improbability and which contains significance for us. The trouble is that as creatures who further our grasp on the world by means of reason, rationality itself assumes significance for us; we prize it and seek after it.

We value our set of rational constructs, our theory of the world, by the extent of its convenience in defining our place in nature – by giving us a place to stand so to speak. As Festinger showed, so tenaciously do we cling to these constructs that we find it easier to entertain simultaneously contradictory beliefs than to relinquish our grasp on our personal reality.

Yet the world view we cling to so tenaciously is no more than an illusion created by reason. The macroscopic level at which we observe events is one which holds a peculiar significance for us as humans. At the atomic level, events occur with a randomness which appears chaotic, and it is only when we observe the statistical effects that appear when billions of atomic events are aggregated that the humanly significant events emerge that science dignifies with the term 'natural law'.

This sea of random events at the borders of existence, the physicists' 'quantum foam', is every bit as dangerous and as attractive as Kant's 'land of truth surrounded by a wide and stormy ocean, the region of illusion'. In any table of random numbers there are long sequences of repeated patterns and other apparent regularities – artifacts of complexity. So at the quantum level there are macroscopic artifacts, and in the human mind, artifacts of reason. These artifacts may be as interesting as a DNA molecule or as useful as the test-tube that contains it, but they are no more substantial than the fog bank that 'seems to the mariner on his voyage of discovery, a new country, and while constantly deluding him with vain hopes, engages him in dangerous adventures, from which he never can desist, and which he never can bring to a termination.'

Physics takes as its most fundamental proposition the idea that individual human experience is one of the great cosmic constants, varying in a predictably mechanical way only in response to external influences: the assumption that if I measure a piece of string with a ruler and find it ten centimetres long, then you too will find it ten centimetres long. This idea finds its expression in the principle of relativity (expounded by Galileo) which says that valid experiments must be repeatable anywhere, anytime by anybody.

This principle obviously rules out the possibility of one experimenter's experience of the world varying from another's in a purely personal way. Not surprisingly, when Rudi Schneider, or Nina Kulagina, or Uri Geller move objects at a distance in laboratory conditions without touching them, the results are simply dismissed as impossible, because they violate the principle of relativity.

In order to come to terms with phenomena of this sort, science must accept that the world may be so constituted that different individuals have different experiences even when their circumstances appear to be identical. To grasp fully the nature of the world, science must be both rational and non-rational in its approach: it must be possible to grasp the unique and the random as well as the uniform and the law-bound.

Most mystical or religious philosophies have embraced the idea that behind or beyond the world of our senses is a transcendental reality, accessible only to the 'superconscious' mind and which is the source of the phenomena that are termed paranormal. On the face of it, such metaphysical notions are quite at odds with the rational tradition of western science – yet paradoxically they represent merely different facets of one and the same world view.

It seems to me that the position in which we actually find ourselves is infinitely more remarkable and infinitely more disquieting. For the phenomenal world of our senses *is* the transcendent reality we inhabit. In the words of Alfred Whitehead:

> Nature gets credit which in truth should be reserved for ourselves, the rose for its scent, the nightingale for his song, and the sun for its radiance. The poets are entirely mistaken. They should address their lyrics to themselves and should turn them into odes of self-congratulation on the excellence of the human mind. Nature is a dull affair, soundless, scentless, colourless, merely the hurrying of material, endlessly, meaninglessly.

Congratulations are due not only to poets but to scientists too, for their ingenuity in bestowing the appearance of objectivity on this very personal world we inhabit. But beyond or behind this world is the mad microcosm of quantum mechanics where, so far, scientific law has been unable to penetrate. Only a non-rational, intuitive science can gain access to this 'endless, meaningless hurrying of material' – a science of the imagination.

To many in the West, this may sound no more than a romantic delusion. But it is simply another way of saying there are important sources of valuable human knowledge quite apart from scientific knowledge and that this non-scientific and even non-rational knowledge can be utilised for purely practical purposes in the organisation and operation of community affairs.

We have an enormously rich and powerful non-scientific cultural heritage on which to draw for support but which has been neglected and out of fashion for decades: philosophy, art, literature, history and even politics in its original, Aristotelian sense of the moral philosophy of the social organism. Those who believe in and practise these arts have too often been marginalised by scientific fundamentalism, while reason has come to dominate our society. Yet, as Blaise Pascal observed, 'the heart has its reasons that reason knows nothing of', and for science – the profession of knowledge – to ignore the reasons of the heart' is to remain in ignorance of a fount of knowledge that is of fundamental importance and practical value.

Postscript

....

CHAPTER FIFTEEN

Frauds, Fakes and Facing Facts

The examples and the research results given in this book are an attempt to illustrate and substantiate my main argument: that there are legitimate areas of scientific research that are being neglected for non-scientific reasons; that a subtle form of scientific censorship is being applied to such research; and that, as a result, important scientific discoveries may be ignored, or even lost to us entirely.

There is one substantial objection that may be raised against the examples and arguments that I have detailed here and it is that, while I have looked in some detail at self-deception as it sometimes occurs in orthodox scientific research, I have made no attempt systematically to investigate and comment on the many cases of deliberate fraud and deception which have plagued science as they plague most human undertakings.

The strength of this objection arises from the fact that it is precisely in the murky areas of alleged anomalous phenomena such as the paranormal that charlatanism is most likely to flourish, and thus practically the entire body of evidence I have brought forward to substantiate such phenomena could be discounted on the grounds of its doubtful authenticity alone.

It can be alleged for example, apparently with some justification, that I have rather gullibly accepted at face value the findings of researchers in very controversial and suspect areas of research, such as extra-sensory perception and psychokinesis, while not raising a finger to point at the undoubted charlatans and con-artists who inhabit the shadowy world in which there is money to be made out of people's willingness to believe impossible things.

Why have I defended the tests on Uri Geller, while omitting to point out that several serious researchers (including Geller's own supporters) have expressed doubts about his performances, and that there is some evidence that he may have cheated on occasions (though not in the laboratory results reported here)?

In this postscript, I wish to explain the specific reasons why I have omitted these questions from my main arguments and of the true place that I believe that they should have in this debate.

In order to show the difficulty that scientific fraud poses to anyone interested in the history and philosophy of science, let me illustrate the point with two contrasting examples from the world of biology: those of Ernst Haeckel and Paul Kammerer.

While researching a previous book on evolutionary biology, *The Facts of Life*,[1] I looked in some detail at these two biologists. Historically, both are considered to be fraudulent researchers, in that both were alleged to have 'improved' the research data that apparently proved the scientific cases they wished to make. It would have given considerable support to my argument then to have recounted that part of their work that involved the alleged frauds. Yet the more I investigated their stories, the more I became convinced that to include either of them would be misleading rather than illuminating. Let me explain why.

Paul Kammerer was an Austrian biologist, born in Vienna in 1880 and who committed suicide in 1926. Throughout most of his life he was a distinguished experimental researcher with an international reputation. *Nature* magazine called his last book 'one of the finest contributions to the theory of evolution which has appeared since Darwin.' Surprisingly, however, Kammerer's work did not support the evolutionary views of Darwin, but on the contrary provides some of the most convincing experimental evidence ever produced of an evolutionary mechanism far more important than the Darwinian mechanism: a mechanism that is at present denied entirely – the inheritance of acquired characteristics. Kammerer's story was brought to a modern audience by Arthur Koestler in his book *The Case of the Midwife Toad*.[2]

Kammerer worked at the prestigious Institute for Experimental Biology in Vienna under Professor Hans Przibram from 1903 until his death. Over several decades he carried out intricate breeding experiments with many generations of animals and plants to try to find evidence that individuals evolve not because of the selection of chance mutations (the Darwinian idea) but because they were in some unknown way able to adapt their physical features to their habitat or way of life.

Kammerer searched the animal and plant kingdoms, both on land and in water, looking for individuals he could breed in the laboratory that might exhibit this kind of evolution. He found many such examples. He bred spotted salamanders on different colour soils and found that over successive generations they changed colour to resemble that of the soil on which they were bred: those bred on yellow soil showed a progressive enlargement of the yellow spots on their bodies until they became predominantly yellow, while those reared on black soil showed a diminution of the yellow spots until they became predominantly black. When the offspring of these genetically modified salamanders were moved to the opposite colour soil to that of their parents, their coloration changed back again.

It is important to appreciate that this kind of genetic evolutionary change is entirely anti-Darwinian in nature. It is an example of *directed* genetic change (although the mechanism that directs it is entirely unknown); a heresy that all Darwinists vehemently deny is possible.

Kammerer found other such examples. He experimented with the sea squirt *Ciona* which lives on the sea bottom and has two siphons or tubes that wave in the sea above it: one taking in water, the other expelling it again. It was already known that if you cut off the sea squirt's siphons it would grow new ones that were longer than the originals. The more often you cut them off, the longer they grew. Kammerer now bred from such artificially elongated specimens and found that the elongated siphons were inherited. This was simple, direct evidence of the inheritance of an acquired characteristic — something that Darwinists believe is impossible in any organism.

Detailed photographs and descriptions of both the sea squirts and the salamanders were published in scientific journals, and the specimens themselves were freely available for inspection by visiting biologists, a steady stream of whom made a trip to the Institute in Vienna from all of Europe's leading universities. It is impossible that these examples were fraudulent unless not only Kammerer but also the other senior staff of one of the world's most prestigious biological institutes were all practising some kind of surgico-taxidermal fraud on a gigantic scale over a twenty-year period and were able to enlist all their skeptical visitors in the conspiracy.

The experiment that led to Kammerer's downfall and suicide was that performed with many generations of the midwife toad, *Alytes obstetricans*. To mate, a male toad clasps the female around the waist and hangs on (for days or weeks) until she spawns her eggs, which he then fertilises. Because the female is slippery in water, the male develops dark coloured 'nuptial pads' on its palms and fingers to give it a grip.

The midwife toad, alone among its kind, mates on land, and thus has no nuptial pads. (Instead of hanging on to the female, the male collects her eggs and carries them about with him — hence the name 'midwife' toad.) Kammerer bred successive generations of the toad in water, instead of on land, and claimed that they eventually developed nuptial pads as an acquired hereditary feature.

Now, a curious feature of this experiment is that the evidence it provides is nothing like as powerful as that from the experiments with the salamanders or the sea squirts. Indeed, even Kammerer himself said, 'In my opinion it is by no means conclusive proof of the inheritance of acquired characters.' The reason is simply that it is perfectly possible that the midwife toad has descended from ordinary toads and that it had lost the nuptial pads at some time in the past. Their reappearance would simply be a genetic throwback.

It would be quite consistent with a Darwinist interpretation to say that the nuptial pads were controlled by a gene, that at some time in the past the gene had somehow been 'switched off' and had now somehow been 'switched on' again. But there is one important way in which this explanation is entirely anti-Darwinian and that is that there is no known mechanism by which the creature's environment or habits could affect the genes contained in its sexual cells; to switch them on or off. Even if the unaccustomed watery environment somehow produced spontaneous nuptial pads, there is no way this character could be inherited by the next generation unless the mutation affected the genes in the creature's germ cells.

Because of the anti-Darwinian implications of all his experiments, Kammerer became a marked man. Darwinists hounded him and his experiments through the scientific press, practically accusing him of fraudulent work from the outset, even though the photographs of his salamanders and sea squirts are clearly genuine. A pronounced anti-Kammerer scientific faction developed, led principally by the English geneticist William Bateson, which mounted frequent attacks on his work. Often, these attacks did not address the scientific issues but attacked Kammerer's integrity. When they did address scientific issues it was to imply that the experiments *must* be erroneous or fraudulent because the inheritance of acquired characteristics is impossible.

On 7 August 1926, *Nature* carried an article by Dr G.K. Noble, curator of reptiles at the American Museum of Natural History. In it Noble asserted that Kammerer's most famous specimen of the mid-wife toad (and by this date, the only remaining specimen preserved from his breeding experiments) was fake. Noble had examined this last preserved example in Vienna and found that instead of possessing true nuptial pads, someone had injected indian ink under the skin.

Noble's article in *Nature* said that, 'It has therefore been established beyond the shadow of a doubt that the only one of Kammerer's modified specimens of *Alytes* now in existence lacks all trace of nuptial pads. The question remains: might not this specimen at one time have possessed them?' Noble then concluded, 'Whether or not the specimen ever possessed them is a matter for conjecture.' Although he did not say so, it was clear that Noble did not believe the nuptial pads had ever existed.

Nature also published a companion piece from Kammerer's friend and chief, Hans Przibram, who stoutly defended him.

> It is clear from the foregoing account [Noble's article] that the only one of Kammerer's experimentally modified *Alytes* still preserved cannot in its present state be regarded as a valid proof of the nuptial pads artificially produced in this species. We must endeavour to decide if the state the specimen is in now agrees with the state at the time of its preservation and before. The specimen is poorly fixed and preserved. Moreover, the epidermis is in several places ready to be shed or even shedding. It is a known fact, as Professor Franz Werner of Vienna asserts, that during repeated handling and shaking, the nuptial asperities get lost easily. The specimen has made the voyage

to England and back again, and it does not look the better for it. Fortunately, there are photographic plates in existence showing the state of the specimen before it left Vienna for Cambridge, and during its stay in England.

Przibram turns a little later in the article to the artificial colouring of the toad's hand.

> While it is possible to come to a probable solution with respect to the [lost spines referred to earlier] we have not been able to elucidate the origin of the black substance. It is clear that it has nothing to do with the black pigment often seen in conjunction with nuptial pads. The only possibility that we can think of is that someone has tried to preserve the aspect of such black nuptial pads in fear of their vanishing by the destruction of the melanin through exposure to the sun in the museum case, by injecting the specimen with Indian ink. Kammerer himself was greatly astonished at the results of the chemical tests, and it ought to be stated that he had been asked and had given his consent to the chemical investigation.

Some six weeks after publication of the report, Kammerer shot himself on a remote mountain path.

Since it proved impossible for the scientific experts closest to the matter at the time to reach any definite conclusion on the question of the fake nuptial pads, it would be folly to attempt to do so seventy years later. A number of theories were floated at the time: that, for instance, Kammerer – or more likely a lab assistant – had innocently attempted to preserve a fading coloration in the only remaining specimen. There were even darker hints of attempts by Darwinists to discredit him – perhaps while the specimen was out of Kammerer's possession and on Darwin's own soil, in England.

Though it is probably impossible to reach any definite conclusion as to the question of fraud, there is at least one important specific conclusion that can be drawn from the Kammerer affair: that, whether or not the tampering with the preserved specimen was deliberate fraud, such a fraud would not have been very effective since the midwife toad did *not* in any case provide important evidence against Darwinism – the experiments with sea squirts and salamanders were far more important in this respect. Yet with the exposure of the fake midwife toad, *all* Kammerer's work became suspect and would in future be rejected as having no scientific value.

To have included an account of Kammerer's experimental evidence regarding sea squirts and salamanders in *The Facts of Life* would have added powerful evidence to my case. Yet had I done so, I would have been accused of gullibly accepting the evidence of a fraudulent experimenter. All the evidence I presented would in turn have been tainted by such an accusation. Although I had in my hand powerful scientific evidence of undoubted probative value that has still not been widely recognised, I felt unable to use it.

The second scientist whose work I investigated, presented me with exactly the opposite problem. Ernst Haeckel was a well-known nineteenth-century biologist (1834–1919) who dedicated his career to constructing the family tree of mankind, and indeed all living things. He performed a service to biology by coining that very handy word 'ecology' and quite a number of other biological terms. Haeckel was both an original researcher and a great populariser. He was also a great controversialist and above all a dedicated devotee of Darwin.

Haeckel became professor of zoology at Jena University in 1862. Four years later he produced the first of a series of major published works all aimed at reconstructing the history of life on earth along Darwinian lines. Haeckel was an accomplished draughtsman and he illustrated most of his works himself; the first Darwinist to attempt drawing family trees for all living organisms.

Perhaps because he wrote in a heroic age of scientific discovery when great theories were commonplace, Haeckel had a tendency to jump to grand conclusions that were far in advance of any real data. He studied the single-celled aquatic organisms called Radiolaria which, under the microscope, show beautifully symmetrical crystalline forms. Haeckel jumped from this finding to the conclusion that the simplest organic forms of life had formed spontaneously from inorganic chemicals in a kind of crystallisation process.

Later he became fascinated by embryology and began to study and to draw embryos of many different creatures. He ultimately conceived what he called the 'biogenetic law' that ontogeny recapitulates phylogeny (the embryo of any species repeats during its development the evolution of its entire phylum or ancestral tribe) an idea that he converted most Darwinists to – including Darwin himself.

Haeckel wrote:

> When we see that at a certain stage the embryos of man and the ape, the dog and the rabbit, the pig and sheep, though recognisable as vertebrates, cannot be distinguished from each other, the fact can only be elucidated by assuming a common parentage . . . I have illustrated this significant fact by a juxtaposition of corresponding stages in the development of different vertebrates in my *Natural History of Creation*.

It is perfectly true that the illustrations Haeckel had drawn to portray this inter-species affinity did indeed show a remarkable similarity: but the explanation was a great deal simpler than that of common ancestry. It was that Haeckel had embellished his drawings – chopping bits off here, sticking them back on there – to make them look similar. Actually the embryos of humans, apes, dogs, rabbits, pigs and sheep are not the same, but differ in crucially important details.

In another book, Haeckel perpetrated an even more bare-face fraud. He illustrated the similarity of the worm-like stage of embryonic development of a dog, a chicken and a turtle by the simple expedient of using the *same* woodcut three times.

This particular fraud was exposed by L. Rutimeyer, professor of zoology and comparative anatomy at the University of Basel, in Switzerland. Many other examples of forgery were found in his work. On one occasion, for example, he added a couple of bones to an illustration of the human backbone, giving thirty-five vertebrae instead of thirty-three, and topped the whole thing off with a tail containing an extra nine vertebrae.

He was eventually charged with fraud by a university court at Jena. Although he admitted altering his drawings, Haeckel escaped being dismissed or even disgraced. In 1908, he defended himself in a Berlin newspaper by writing:

> To cut short this unsavoury dispute, I begin at once with my contrite confession that a small fraction of my drawings of embryos (perhaps six or eight per cent) are in a sense falsified – all those, namely, for which the present material of observation is so

incomplete or insufficient as to compel us, when we come to prepare a continuous chain of evolutionary stages, to fill up the gaps by hypotheses, and to reconstruct the missing links by comparative synthesis . . . After this compromising confession of 'forgery' I should be obliged to consider myself 'condemned and annihilated' if I had not the consolation of seeing side by side with me in the prisoner's dock hundreds of fellow-culprits, among many of the most trusted observers and esteemed biologists. The great majority of all of the diagrams in the best biological text books, treatises and journals would incur in the same degree the charge of 'forgery', for all of them are inexact, and are more or less doctored, schematized and constructed.[3]

Once again, it would have strengthened my case to have included an account of Haeckel's fake drawings simply to show the extent to which Darwinists have deluded themselves and others in the steadfast belief that they are merely filling in the gaps in a theory that is 'obviously' true. I decided not to include such an account because it would have laid me open to the charge of attacking Haeckel's theories by the disreputable means of discrediting the man; a debating trick that I detest more than any other and the one most frequently employed against anti-Darwinists, like Paul Kammerer. I wanted to show by fair means that Darwinism was not founded in evidence.

Haeckel was not a con-man, setting out to swindle or dupe an innocent public. He was a dedicated professional scientist, more than usually strongly committed to a scientific theory; a theory he believed in so passionately that he saw nothing wrong with assisting it by improving the data, even if that meant making a few slight alterations, in much the same way that someone might well have decided to preserve or bring out a nuptial pad in a specimen of *Alytes* by injecting a little indian ink to emphasise a character they believed to be already present.

The key issue here is that anyone who thinks Haeckel's actions are 'unscientific' has not properly understood the significance of the earlier part of this book (nor Haeckel's impassioned plea in self-defence quoted above). Science does not proceed by fitting perfect data to perfect theories. Science proceeds by ignoring anomalies, in the hope that further discoveries will enable them to be incorporated into the elegant theory espoused by the current paradigm.

And even when scientists do have good grounds for believing something to be true, they are often not the right grounds. For example, Hipparchus and Aristotle believed the Earth to be spherical because they had observed the Sun and the Moon to be so shaped and inferred by analogy that the Earth must be too. This is not actually a sound scientific basis for believing in a spherical Earth – it is merely a reasonable view. Like all scientific beliefs it has been arrived at inductively and is thus liable to the fallacy of induction. Haeckel's grounds for doctoring his drawings were probably just as reasonable as Aristotle's grounds for depicting the Earth as spherical, by the canons of the contemporary paradigm: he merely had the misfortune to be wrong. His crime was pig-headedness, not fraud.

It seems to me that to indulge in unsubstantiated accusations of fraud is completely unproductive and may well be unjust. I believe that the only attitude to the problem worth adopting is a scientific one – if there is evidence of fraud, to present that evidence; if there is no concrete evidence of fraud, to pass over it in silence. In the case of Geller, there is no concrete evidence of fraud, merely suspicion. This is not to say that we should not be on our guard. Both Arthur Koestler and Arthur C. Clarke, who were present at sessions with Geller, are said to have been undecided or doubtful about Geller and a little suspicious of him, although neither publicly gave any reason for such suspicions.

In the absence of any concrete data, it seems to be merely muddying the already unclear waters to indulge in speculation about possible fraud. A better approach seems to me to be watchful, to sift the evidence very carefully, to reject evidence that does not smell right, as I did with Kammerer and Haeckel but to face up squarely to the concrete data that remain; not to sweep them under the carpet with a blanket accusation of 'fraud' or, even more damaging, 'possible fraud'.

That is why I have not reported any of Geller's so-called 'telepathy' experiments, his claims to have 'teleported' objects at will, or indeed his claims to be telepathically in contact with UFOs, but have restricted myself to independent laboratory studies that have been videotaped or for which the instrumental measurements have been chart recorded and are available for inspection.

The procedure I have tried to adopt throughout this book is that recommended by Dr Peter Sturrock and described in Chapter 11. Where a critical evaluation of results is called for I have assigned probability values to the data, rather than accepting or rejecting it. At the same time, I have excluded the probability values of complete certainty and complete disbelief for the reasons given by Dr Sturrock, that once such a value has been assigned, it cannot ever be changed by new information – it amounts to a closed mind.

In the case of phenomena about which most people are naturally suspicious, like psychokinesis and ESP, I have tried to allow the data to speak for itself. Even the odds of 1 in 10^{35} against Radin and Nelson's results being obtained by chance do not allow us to say we know for *certain* that psychokinesis is possible, but they do allow us to say that Geller and others might be able to influence electronic instruments in an unknown way, and that we should look more carefully to find out.

The hundreds of such examples in this book show that there are major anomalous phenomena taking place that orthodox science is ignoring yet which are valid subjects of study, crying out for research, and in many cases promising radical revisions of our understanding of the nature of the world.

It may very well be the case that some of the examples I have given in good faith will turn out to have normal explanations, including in some cases deliberate fraud. If any particular case is shown to be flawed in such a way, then many orthodox scientists will no doubt heave a sigh of relief and return with a clear conscience to their bad old ways, safe in the belief that my entire thesis has been shown to be flawed. The fact remains, however, that if only one tenth – or, indeed, one hundredth – of the examples given here are valid, then it means that orthodox science *must* be prepared to put itself out enough to seek answers to these anomalies, or it must be prepared to abdicate its title as the sovereign means of acquiring knowledge.

To those reductionist scientists who remain skeptical of the validity of the anomalous research data given in this book, and who believe that science is at the end, or close to the end, of its quest for a globally consistent explanation of the nature of the world, I offer the challenge of the following thirteen questions.

- Radin and Nelson have demonstrated psychokinetic effects in controlled laboratory conditions with odds of 1 in 10^{35} that their results arose by chance. What is the mechanism of this effect?[4]
- Excess energy has been reproduced experimentally from Fleischmann-Pons cells in 92 research organisations in ten countries. What is the explanation of 'cold fusion'?[5]
- What is the explanation of the non-random mutation of E. coli produced experimentally by Cairns and replicated by Hall?[6]
- What are the implications for living organisms of the discovery by Piccardi that chemical reactions in water are influenced by electromagnetic radiation?[7]
- What are the implications of the experimental findings of Gurwitsch replicated by Shchurin that some organisms under some circumstances appear to be able to communicate information at ultraviolet or shorter wavelengths?[8]
- Replicated trials have established that acupressure and acupuncture can produce repeatable physiological effects; how?[9] If there are energy meridians in the human body, how are they mapped? What is their source? Why are we unable to detect them by conventional methods?
- Several laboratories have shown that if part of a leaf is removed, the entire original shape is still visible by Kirlian (corona discharge) photography for some time after. What is the explanation of this persistence? What is being photographed?[10]
- Emotional states such as bereavement have been reliably shown experimentally to affect immune system functions; how? And how far?[11]
- Eysenck has shown that personality variables are more predictive of death through heart disease and cancer than smoking. If emotional stress kills; how?[12]
- How exactly does the placebo effect work? If the answer is 'suggestion' then how *exactly* does suggestion work? What are its limits?
- Penfield has demonstrated experimentally that brain functions such as pattern recognition and speech are called upon alternately by some higher function which he calls merely the 'mind'. Where

does this higher mental function reside? What is its connection with the brain? What are its capabilities?[13]

• The statistical study of Michel Gauquelin that established a correlation between certain professions and the positions of certain planets at birth has now been replicated several times. What is the explanation of this correlation?[14]

• What is the exact nature of any so-called field phenomenon — that is, any action-at-a-distance? *Exactly* how do like magnetic poles repel, for instance?

If you believe that current science adequately answers these questions, then do not trouble your mind any further. If you feel the shadow of a doubt when looking through them, then ask yourself just where new answers to these questions might lead science.

Notes

CHAPTER 1

Too Wonderful to be True

1 Reichenbach, Karl von, 1850 (trs. William Gregory), *Researches on Magnetism, Electricity etc.*, Taylor, Walton & Maberly, London.
2 Fleischmann, M., and Pons, S., 1989, *Electrochemically induced nuclear fusion of deuterium*, in *Journal of Electroanalytical Chemistry*, Vol. 261, pp. 301–8; and Vol. 263, pp. 187–8.
3 See Chapter 3 for detailed references.
4 See Chapter 6 for detailed references.
5 Popper, Karl, 1977 edn., *The Open Society and its Enemies*, Routledge & Kegan Paul, London.
6 *Britain 1993*, HMSO, pp. 364 and 380.
7 See Chapter 7 for detailed references.

CHAPTER 2

A Completely Idiotic Idea

1 Kelly, Fred C., 1944, *The Wright Brothers*, Harrap, London.
2 *The Independent*, 22 October 1903.
3 Kelly, *The Wright Brothers*.
4 *Scientific American*, 13 January 1906.
5 Talbot, Frederick A., 1912, *Steamship Conquest of the World*, William Heinemann, London, pp. 44–6.
6 Taylor, Gordon Rattray (ed.), 1982, *The Inventions That Changed the World*, Reader's Digest, London.
7 Pollen, Anthony, 1980, *The Great Gunnery Scandal*, Collins, London.
8 Macintyre, Donald, 1957, *Jutland*, Evans Brothers, London.
9 Conot, Robert, 1992, Thomas A. Edison: *A Streak of Luck*, Da Capo, New York.

10 Cooper, Bryan, 1967, *The Ironclads of Cambrai*, Souvenir Press, London, p. 28.

11 Foley, John, 1981, *The Boilerplate War*, W.H. Allen, London, p. 4.

12 *The Times*, 28 January 1926.

13 Norman, Bruce, 1984, *Here's Looking at You*, BBC, London, p. 37.

14 Taylor, *Inventions*.

CHAPTER 3

Sunbeams from Cucumbers

1 Fleischmann, M. and Pons, S., 1989, *Electrochemically induced nuclear fusion of deuterium*, in *Journal of Electroanalytical Chemistry*, Vol. 261, pp. 301–8; and Vol. 263, pp. 187–8.

2 *New Scientist*, 22 April 1989, p. 27.

3 *New Scientist*, 29 April 1989, pp. 22–3.

4 *The Independent*, 20 April 1989.

5 *New Scientist*, 22 April 1989, p. 27.

6 *Daily Telegraph*, 1 April 1989.

7 *New Scientist*, 6 May 1989, p. 26.

8 *Daily Telegraph*, 2 May 1989, p. 1.

9 *Newsletter of the National Association of Sciencewriters*, Vol. 39, no. 3, Fall 1991, pp. 24–5.

10 Swartz, Mitchell R., 1992, *Reexamination of key cold fusion experiment by the MIT Plasma Fusion Laboratory*, in *Fusion Facts* (newsletter of Fusion Information Center, University of Utah), Vol. 4, no. 2, pp. 27–41.

11 *Newsletter of the National Association of Sciencewriters*, Vol. 39, no. 3, Fall 1991, pp. 24–5.

12 Swartz, 1992, *Reexamination of key cold fusion experiment*, p. 320.

13 Albagli, D., *et al.*, 1990, *Measurement and analysis of neutron and gamma-ray emission rates etc.*, in *Journal of Fusion Energy*, Vol. 9, p. 133. The conclusion should be compared with Luckhardt, S.C., May 1992, *Technical appendix to D. Albagli, et al., J. Fusion Energy*, 1990, *Calorimetry error analysis*, MIT Report PFC/RR-92-7.

14 *Cold Fusion 1992: Basic Facts* (Newsletter published by Clustron Sciences Corporation, Vienna, Va.). See also Mallove, Eugene, 1991, *Fire From Ice*, John Wiley, New York.

15 *Nature* editorial, 29 March 1990, p. 365.

16 *Daily Telegraph*, 15 May 1989, p. 12.

17 *Daily Telegraph*, 12 June 1989.

18 *Daily Telegraph*, 1 April 1989.

19 *Daily Telegraph*, 18 April 1989.

20 Huizenga, John R., 1992, *Cold Fusion: Scientific Fiasco of the Century*, University of Rochester Press.

CHAPTER 4

The Gates of Unreason

1 Hasted, John, 1981, *The Metal Benders*, Routledge & Kegan Paul, London.

2 Piccardi, Giorgio, 1960, Exposé introductif, *Symposium Intern. sur les Rel. Phen. Sol et Terre*, Presses Académiques Européennes, Brussels.

Piccardi, Giorgio, 1962, *The Chemical Basis of Medical Climatology*, C.C. Thomas, Springfield, Ill.

3 Capel-Boute, C., 1960, *Observations sur les tests chimiques de Piccardi*, Presses Académiques Européennes, Brussels.

4 Fisher, W., Sturdy, G., Ryan, M., and Pugh, R., 1969, *Some laboratory studies of fluctuating phenomena*, in Gauquelin, *The Cosmic Clocks*, Peter Owen, London.

5 Hasted, John, 1981, *The Metal Benders*, Routledge & Kegan Paul, London.

6 Puthoff, Harold, and Targ, Russell, 1974, *Information transmission under condition of sensory shielding*, SRI Report.

7 Taylor, John, 1976, *Superminds*, Pan, London.

8 Taylor, John, 1980, *Science and the Supernatural*, Temple Smith, London.

9 Eysenck, Hans J., and Sargent, Carl, 1982, *Explaining the Unexplained*, Weidenfeld & Nicolson, London.

10 McCreery, Charles, 1967, *Science, Philosophy and ESP*, Faber & Faber, London.

11 Ibid.

12 Fox, Edward, 1992, *Believe it or not*, in *The Independent Magazine*, 14 November, pp. 45–52.

13 Eysenck and Sargent, *Explaining the Unexplained*.

14 Broughton, Richard, 1992, *Parapsychology: The Controversial Science*, Random Century, London.

15 Radin, Dean I., and Nelson, Roger D.,1989, *Consciousness-related effects in random physical systems*, in *Foundations of Physics*, Vol. 19, pp. 1499–1514.

CHAPTER 5

Animal Magnetism

1 Gurwitsch, Alexander, 1937, *Mitogenic analysis of the excitation of the nervous system*, North Holland, Amsterdam.

2 Hollaender, A., and Claus, W.D., 1935, *Journal of the Optical Society of America*, Vol. 25, p. 270.

3 Hall, Robert N. (ed.), 1989, *Pathological science*, in *Physics Today*, October, pp. 36–48.

4 Tompkins, Peter, and Bird, Christopher, 1975 edn., *The Secret Life of Plants*, Penguin, London, pp. 177–9.

5 See Chapter 4 for detailed references.

6 See Chapter 4 for detailed references.

7 Podmore, Frank, 1909, *Mesmerism and Christian Science*, Methuen, London.

8 Ibid.

9 Reichenbach, Karl von, 1850 (trs. William Gregory), *Researches on Magnetism, Electricity etc.*, Taylor, Walton & Maberly, London.

10 Kilner, Walter J., 1975 edn., *The Human Atmosphere*, Samuel Weisner, New York.

11 Reich, Wilhelm, 1971, *Selected Writings*, Vision Press, London.

12 Kirlian, Semyon D., and Kirlian, Valentina H., 1968, *Investigation of Biological Objects in High Frequency Electrical Fields etc.*, Alma Ata, USSR.

13 Moss, Thelma, and Johnson, K., 1972, *Radiation Field Photography*, in *Physics*, July.

14 Taylor, John, 1980, *Science and the Supernatural*, Temple Smith, London.

15 Oldfield, Harry, and Coghill, Roger, 1988, *The Dark Side of the Brain*, Element, Dorset.

16 Ibid.

CHAPTER 6

A Case of Ill Treatment

1 Bartrop, R.W., *et al.*, 1977, *Depressed lymphocyte function after bereavement*, in *Lancet*, 16 April, pp. 834–6.

2 Bower, Bruce, 1991, *Questions of mind over immunity*, in *Science News*, Vol. 139, pp. 216–17.

3 Ibid.

4 Bartrop, R.W., *et al.*, 1977, *Depressed lymphocyte function*.

5 Eysenck, H.J., 1988, *Personality, stress and cancer: prediction and prophylaxis*, in *British Journal of Medical Psychology*, Vol. 61, pp. 57–75.

6 Spiegel, David, *et al.*, 1989, *Effects of psychosocial treatment on survival of patients with metastatic breast cancer*, in *Lancet*, 14 October, pp. 888–91.

7 HMSO, 1991, *The Health of the Nation*. Government Green Paper on the future of the Health Service.

8 British Holistic Medical Association (BHMA), October 1991, *A Response to the Government's Green Paper*, BHMA, London.

9 Nixon, P.G.F., 1991, *Cardiovascular Health Promotion and Rehabilitation*, in *BHMA, a response*.

10 Dundee, J.W., *et al.*, 1988, *P6 acupressure reduces morning sickness*, in *Journal of the Royal Society of Medicine*, Vol. 81, pp. 456–7.

11 Dundee, J.W., *et al.*, 1987, *P6 Acupuncture: an effective non-toxic anti-emetic in cancer chemotherapy*, in *British Journal of Anaesthesia*, Vol. 59, p. 1322.

12 Fry, E.N.S., 1986, *Acupressure and postoperative vomiting*, in *Anaesthesia*, Vol. 41, pp. 661–2.

13 Ornish, Dean, *et al.*, 1990, *The lifestyle heart trial*, in *Lancet*, Vol. 336, pp. 129–33.

14 Conway, A.V., 1986, *Cancer and the mind: a role for hypnosis?*, in *Holistic Medicine*, Vol. 1, pp. 43–5.

15 Conway, A.V., 1988, *The research game: a view from the field*, in *Complementary Medical Research*, Vol. 3, No. 8, pp. 29–36.

16 O'Regan, B., 1987, *Healing, remission and miracle cures*, Special Report of the Institute of Noetic Sciences, Vol. 3, p. 14.

17 Achterberg, J., *et al.*, 1977, *Psychology of the exceptional cancer patient etc.*, in *Psychotherapy Theory, Research and Practice*, Vol. 14, pp. 416–22.

18 Rees, W.D., *et al.*, 1967, *Mortality of bereavement*, in *British Medical Journal*, Vol. 4, pp. 13–16.

19 Bartrop, R.W., *Depressed lymphocyte function*.

20 Schliefer, S.J., *et al.*, 1983, *Suppression of lymphocyte stimulation following bereavement*, in *Journal of the American Medical Association*, Vol. 250, pp. 374–7.

21 Ghanta, V.J., *et al.*, 1985, *Neural and environmental influences on neoplasia and conditioning of NK activity*, in *Journal of Immunology*, Vol. 135, pp. 848–52.

22 Simonton, O.C., *et al.*, 1980, *Getting Well Again*, Bantam, London.

23 Haines, A.P., *et al.*, 1987, *Phobic anxiety and ischaemic heart disease*, in *British Medical Journal*, Vol. 295, pp. 297–9.

24 Conway, A.V., 1988, *The research game*.

25 HMSO, 1993, *Government Review of R & D 1992*.

26 Quoted in *Time*, 4 November 1991, p. 75.

27 BHMA, 1991, *A response*.

CHAPTER 7

Forbidden Fields

1 *New Scientist*, 18 August 1988, p. 19.

2 Broughton, Richard, 1991, *Parapsychology*, Random Century, London.

3 Milton, Richard, 1993 edn., *The Facts of Life: Shattering the Myths of Darwinism*, Corgi, London.

4 *Britain 1993*, HMSO, pp. 364 and 380.

5 Ibid.

6 Ibid.

7 British Holistic Medical Association (BHMA), October 1991, *A Response to the Government's Green Paper*, BHMA, London.

8 Ibid.

9 Horrobin, David, 1982, *In praise of non-experts*, in *New Scientist*, 24 June 1982, pp. 842–4.

10 Bailar, John C., and Smith, Elaine M., 1986, *Progress against cancer?*, in *New England Journal of Medicine*, 8 May, pp. 1226–32.

11 *Cancer Patient Survival: What Progress has Been Made?*, 1987, US General Accounting Office.

12 Sharpe, Robert, 1991, *The war against cancer*, in *Outrage*, the Journal of Animal Aid, Jun/Jul 1991, no. 74, pp. 7–10.

13 Ibid., p 10.

14 Smith, Richard, 1988, *Research begins at forty*, in *New Scientist*, 30 June 1988, pp. 54–8.

15 Ibid.

CHAPTER 8

Calling a Spade a Spade

1 Bruner, J.S., and Postman, Leo, 1949, *On the perception of incongruity: a paradigm*, in *Journal of Personality*, Vol. 18, pp. 206–23.

2 Kuhn, Thomas S., 1970 edn., *The Structure of Scientific Revolutions*, University of Chicago Press, p. 64.

3 Festinger, Leon, 1962 edn., *A Theory of Cognitive Dissonance*, Stanford University Press, Calif.

4 Schacter, S., 1951, *Deviation, rejection and communication*, in the *Journal of Abnormal and Social Psychology*, Vol. 46, pp. 129–44.

5 Festinger, *A Theory of Cognitive Dissonance*.

6 Kuhn, *The Structure of Scientific Revolutions*.

7 Thompson, Sylvanus P., 1910, *The Life of Sir William Thomson, Baron Kelvin of Largs*, London.

8 Kuhn, *The Structure of Scientific Revolutions*.

9 Hahn, Otto, and Strassman, Fritz, 1939, *Uber den Nachweis und das Verhalten der bei der Bestrahlung des Urans mittels Neutronen entstehended Erdalkalmetalle*, in *Die Naturwissenschaften*, Vol. 27, p. 15.

10 Stratton, George M., 1897, *Vision without inversion of the retinal image*, in the *Psychological Review*, Vol. 4, pp. 341–60.

11 Kuhn, *The Structure of Scientific Revolutions*.

CHAPTER 9

The Research Game

1 Dixon, Norman, 1976, *On the Psychology of Military Incompetence*, Jonathan Cape, London.

2 Adorno, T.W., *et al.*, 1950, *The Authoritarian Personality*, Harper, New York.

3 Eysenck, H.J., 1970, *The Structure of Human Personality*, Methuen, London.

4 Dixon, *On the Psychology of Military Incompetence*.

5 Ibid.

6 Conway, A.V., 1988, *The research game: a view from the field*, in *Complementary Medical Research*, Vol. 3, No. 8, pp. 29–36.

7 Horrobin, David, 1982, *In praise of non-experts*, in *New Scientist*, 24 June 1982, pp. 842–4.

8 Maier, S.F., and Seligman, M.E.P., 1976, *Learned helplessness: theory and evidence*, in *Journal of Experimental Psychology*, Vol. 105, pp. 2–46.

9 Lewis, C.S., 1983 edn., *That Hideous Strength*, Pan, London.

10 Goodchild, Peter, 1980, *J. Robert Oppenheimer*, BBC Books, London, p. 151.

11 Ibid., p. 152.

12 Watson, James D., 1968, *The Double Helix*, Weidenfeld & Nicolson, London, p. 116.

13 Ibid., pp. 159–63.

CHAPTER 10

Guardians of the Gate

1 *Nature*, 1981, Vol. 293, pp. 245–6.

2 Velikovsky, Immanuel, 1973 edn., *Worlds in Collision*, Sphere, London.

3 *American Behavioural Scientist*, September 1968.

4 De Grazia, Alfred, Juergens, R.E. and Stecchini, L.C. 1966, *The Velikovsky Affair*, Sidgwick & Jackson, London.

5 Juergens, Ralph, 1968, *Minds in chaos*, in *American Behavioural Scientist*, September 1968.

6 *New York Times*, 11 June 1950.

7 *Missiles and Rockets*, 18 January 1965.

8 De Grazia, *The Velikovsky Affair*, p. 188.

9 Ibid., p. 190.

10 Lafleur, Laurence J., 1951, *Cranks and Scientists*, in *Scientific Monthly*, November, pp. 284–90.

11 Puthoff, Harold, and Targ, Russell, 1974, *Information transmission under condition of sensory shielding*, SRI Report, and *Nature*, 18 October 1974.

12 Evans, Christopher, 1973, *Cults of Unreason*, George G. Harrap, London.

13 Festinger, Leon, 1962 edn., *A Theory of Cognitive Dissonance*, Stanford University Press, Calif.

14 Randi, James, 1975 edn., *The Magic of Uri Geller*, Ballantine, New York.

15 Ibid.

16 *Nature*, 18 October 1974.

17 Clare, Anthony, and Thompson, Sally, 1981, *Let's Talk About Me*, BBC Books, London.

18 Taylor, John, 1976, *Superminds*, Pan, London.

19 Taylor, John, 1980, *Science and the Supernatural*, Temple Smith, London.

20 Price, George R., 1955, *Science and the Supernatural*, in *Science*, Vol. 122, pp. 359–67. Compare with Price, George R., 1972, *Apology to Rhine and Soal*, in *Science*, Vol. 175, p. 359.

21 Kurtz, Paul, 1976, *Committee to Scientifically Investigate Claims of the Paranormal and Other Phenomena*, in *The Humanist*, May/June 1976, p. 28.

22 Sturrock, Peter, 1988, *Brave New Heresies*, in *New Scientist*, 24/31 December 1988, pp. 49–51.

23 Hansel, C.E.M., 1980, *ESP and Parapsychology: A Critical Re-evaluation*, Prometheus, Buffalo.

24 Gauquelin, Michel, 1969, *The Cosmic Clocks*, Peter Owen, London.

25 Melton, J. Gordon, *et al.*, 1990, *Skeptics and the New Age*, in *New Age Encyclopaedia*, Gail Research, Detroit.

26 Randi, James, 1975 edn., *The Magic of Uri Geller*, Ballantine Books, New York.

27 Broughton, Richard, 1991, *Parapsychology: The Controversial Science*, Random Century, London.

CHAPTER 11

A Trout in the Milk

1 Trench, Brinsley le Poer, 1974, *Secret of the Ages*, Souvenir Press, London.

2 Hall, Robert N., (ed.), 1989, *Pathological science*, in *Physics Today*, October 1989, pp. 36–48.

3 Ibid.

4 Ibid.

5 Davis, B., and Barnes, A.H., 1935, *Physical Review*, Vol. 37, p. 1368.

6 Tompkins, Peter, and Bird, Christopher, 1975 edn., *The Secret Life of Plants*, Penguin, London, pp. 177–9.

7 Latimer, W.M., and Young, H.A., 1933, *The isotopes of hydrogen by the magneto-optic method*, in *Physical Review*, Vol. 44, p. 690.

8 McGhee, J.L., and Lawrentz, M., 1932, *Test for element 87 by the use of Allison's magneto-optic apparatus*, in *Journal of the American Chemical Society*, Vol. 54, p. 405.

9 Ball, T.R., 1935, *Observations on the Allison Magneto-Optic Apparatus*, in *Physical Review*, Vol. 47, p. 548.

10 *Science Fiction?*, television film written and produced by Hilary Lawson and broadcast in the *Horizon* series by BBC2 in 1986.

11 *Daily Telegraph*, 15 May 1989, p. 12.

12 Lafleur, Laurence J., 1951, *Cranks and scientists*, in *The Scientific Monthly*, November 1951, pp. 284–90.

13 Sturrock, Peter, 1988, *Brave New Heresies*, in *New Scientist*, 24/31 December 1988, pp. 49–51.

CHAPTER 12

Too Insensitive to Confirm

1 Kelly, Fred C., 1944, *The Wright Brothers*, Harrap, London.

2 Medawar, Sir Peter, 1985, *The Limits of Science*, Oxford University Press.

3 Miller, Russell, *Resurrecting Lorenzo*, in *Sunday Times Magazine*, 20 December 1992, pp. 38–44.

4 *Daily Telegraph*, 1 April 1989.

5 Popper, Karl, 1963, *Conjectures and Refutations*, London, Routledge & Kegan Paul, p. 34.

6 *Science Fiction?*, television film written and produced by Hilary Lawson and broadcast in the *Horizon* series by BBC2 in 1986.
7 Schiff, L.I., 1961, *A Report on the NASA conference on experimental tests of relativity theories*, in *Physics Today*, Vol. 14, pp. 42–8.

CHAPTER 13

The Future that Failed

1 Medawar, Sir Peter, 1985, *The Limits of Science*, Oxford University Press.
2 Miller, Henry W., 1930, *The Paris Gun*, George G. Harrap, London.
3 Von Braun, Wernher, 1971 edn., *Space Frontier*, London.
4 Laurie, Peter, 1972, *Beneath the City Streets*, Penguin, London.
5 Carson, Rachel, 1972 edn., *Silent Spring*, Penguin, London.
6 Kant, Immanuel, 1969 edn., *Critique of Pure Reason*, J.M. Dent, London.
7 Crick, Francis, 1966, *Of Molecules and Men*, University of Washington Press, Seattle.

CHAPTER 14

A Methodical Madness

1 Penfield, Wilder, 1975, *The Mystery of the Mind: a Critical Study of Consciousness and the Human Brain*, Princeton University Press.
2 Milton, Richard, 1993 edn., *The Facts of Life: Shattering the Myths of Darwinism*, Corgi, London.
3 Thornton, E.M., 1983, *Freud and Cocaine*, Blond and Briggs, London.
4 Koestler, Arthur, 1971 edn., *The Ghost in the Machine*. Pan, London.
5 Bruner, J.S., and Postman, Leo, 1949, *On the perception of incongruity: a paradigm*, in *Journal of Personality*, Vol. 18, pp. 206–23.
6 Festinger, Leon, 1962 edn., *A Theory of Cognitive Dissonance*, Stanford University Press, Calif.
7 Piaget, Jean, 1967, *Language and Thought of the Child*, Routledge & Kegan Paul, London.

CHAPTER 15

Frauds, Fakes and Facing Facts

1 Milton, Richard, 1993 edn., *The Facts of Life: Shattering the Myths of Darwinism*, Corgi, London.
2 Koestler, Arthur, 1974 edn., *The Case of the Midwife Toad*, Pan, London.

3 Assmuth, J., 1918, *Haeckel's Frauds and Forgeries*, London.

4 Radin, Dean I., and Nelson, Roger D.,1989, *Consciousness-related effects in random physical systems*, in *Foundations of Physics*, Vol. 19, pp. 1499–514.

5 *Cold Fusion 1992: Basic Facts* Newsletter published by Clustron Sciences Corporation, Vienna, Va. See also Mallove, Eugene, 1991, *Fire From Ice*, John Wiley, New York.

6 Cairns, J., Overbaugh, J., and Miller, S., 1988, *The origin of mutants*, in *Nature*, Vol. 335, pp. 142–5.

Hall, Barry G., Sept. 1990, *Spontaneous point mutations that occur more often when advantageous than when neutral*, in *Genetics*, Vol. 126, pp. 5–16.

7 Piccardi, Giorgio, 1960, Exposé introdutif, *Symposium Intern. sur les Rel. Phen. Sol et Terre*, Presses Académiques Européennes, Brussels.

Piccardi, Giorgio, 1962, *The Chemical Basis of Medical Climatology*, Thomas, Springfield, Ill.

Capel-Boute, C., 1960, *Observations sur les tests chimiques de Piccardi*, Presses Académiques Européennes, Brussels.

Fisher, W., Sturdy, G., Ryan, M., and Pugh, R., 1969, *Some Laboratory Studies of Fluctuating Phenomena*, in Gauquelin, *The Cosmic Clocks*, Peter Owen, London.

8 Gurwitsch, Alexander, 1937, *Mitogenic analysis of the excitation of the nervous system*, North Holland, Amsterdam.

Tompkins, Peter, and Bird, Christopher, 1975 edn., *The Secret Life of Plants*, Penguin, London, pp. 177–9.

9 Dundee, J.W., *et al.*, 1988, *P6 acupressure reduces morning sickness*, in *Journal of the Royal Society of Medicine*, Vol. 81, pp. 456–7.

Dundee, J.W., *et al.*, 1987, *P6 Acupuncture: an effective non-toxic anti-emetic in cancer chemotherapy*, in *British Journal of Anaesthesia*, Vol. 59, p. 1322.

Fry, E.N.S., 1986, *Acupressure and postoperative vomiting*, in *Anaesthesia*, Vol. 41, pp. 661–2.

10 Kirlian, Semyon D., and Kirlian, Valentina H., 1968, *Investigation of Biological Objects in High Frequency Electrical Fields etc.*, Alma Ata, USSR.

Moss, Thelma, and Johnson, K., 1972, *Radiation Field Photography*, in July *Physics*.

Oldfield, Harry, and Coghill, Roger, 1988, *The Dark Side of the Brain*, Element, Dorset.

11 Bartrop, R.W., *et al.*, 1977, *Depressed lymphocyte function after bereavement*, in *Lancet*, 16 April 1977, pp. 834–6.

Schliefer, S.J., *et al.*, 1983, *Suppression of lymphocyte stimulation following bereavement*, in *Journal of the American Medical Association*, Vol. 250, pp. 374–7.

12 Eysenck, H.J., 1988, *Personality, stress and cancer: prediction and prophylaxis*, in *British Journal of Medical Psychology*, Vol. 61, pp. 57–75.

13 Penfield, Wilder, 1975, *The Mystery of the Mind: a Critical Study of Consciousness and the Human Brain*, Princeton University Press.

14 Gauquelin, Michel, 1969, *The Cosmic Clocks*, Peter Owen, London.

Bibliography

ACHTERBERG, J., *et al.*, 1977, *Psychology of the exceptional cancer patient etc.*, in *Psychotherapy Theory, Research and Practice*, Vol. 14, pp. 416–22.

ADORNO, T.W., *et al.*, 1950, *The Authoritarian Personality*, Harper, New York.

ALBAGLI, D., *et al.*, 1990, *Measurement and Analysis of Neutron and Gamma-Ray Emission Rates etc.*, in *Journal of Fusion Energy*, Vol. 9, p. 133.

ASSMUTH, J., 1918, *Haeckel's Frauds and Forgeries*, London.

BAILAR, John C., and Smith, Elaine M., 1986, *Progress against Cancer?*, in *The New England Journal of Medicine*, 8 May 1986, pp. 1226–32.

BALL, T.R., 1935, *Observations on the Allison Magneto-Optic Apparatus*, in *Physical Review*, Vol. 47, p. 548.

BARTROP, R.W., *et al.*, 1977, *Depressed lymphocyte function after bereavement*, in *Lancet*, 16 April 1977, pp. 834–6.

BERNE, Eric, 1975 edn., *Games People Play*, Penguin Books, London.

BOHM, David, 1980, *Wholeness and the Implicate Order*, Routledge & Kegan Paul, London.

BOWER, Bruce, 1991, *Questions of Mind Over Immunity*, in *Science News*, Vol. 139, pp. 216–17.

British Holistic Medical Association, October 1991, *A Response to the Government's Green Paper*, BHMA. London.

BROUGHTON, Richard, 1991, 1992 edn., *Parapsychology: the Controversial Science*, Random Century, London.

BRUNER, J.S., and Postman, Leo, 1949, *On the Perception of Incongruity: A Paradigm*, in the *journal of Personality*, Vol. 18, pp. 206–23.

CAIRNS, J., Overbaugh, J., and, Miller, S., 1988, *The origin of mutants*, in *Nature*, Vol. 335, pp. 142–5.

CARSON, Rachel, 1972 edn., *Silent Spring*, Penguin Books, London.

CHALMERS, T.W., 1949, *Historic Researches*, London.

CLARE, Anthony, and Thompson, Sally, 1981, *Let's Talk About Me*, BBC Books, London.

COMBS, Harry B., and Martin, Caidin, 1980, *Kill Devil Hill*, Secker and Warburg, London.

CONOT, Robert, 1992, Thomas A. Edison: *A Streak of Luck*, DaCapo, New York.

CONWAY, A.V., 1986, *Cancer and the mind: a role for hypnosis?*, in *Holistic Medicine*, Vol. 1, pp. 43–5.

CONWAY, A.V., 1988, *The research game: a view from the field*, in *Complementary Medical Research*, Vol. 3, no. 8, pp. 29–36.

COOPER, Bryan, 1967, *The Ironclads of Cambrai*, Souvenir Press, London.

CRICK, Francis, 1966, *Of Molecules and Men*, University of Washington Press, Seattle.

DARROW, K.K., 1940, *Nuclear fission*, in *Bell System Technical Journal*, Vol. 19, pp. 267–89.

DAVIS, B., and Barnes, A.H., 1935, in *Physical Review*, Vol. 37, p. 1368.

DE GRAZIA, Alfred, Juergens, R.E., and Stecchini, L.C., 1966, *The Velikovsky Affair*, Sidgwick and Jackson, London.

DIXON, Norman, 1976, *On the Psychology of Military Incompetence*, Jonathan Cape, London.

DUNDEE, J.W., *et al.*, 1987, *P6 Acupuncture: an effective non-toxic anti-emetic in cancer chemotherapy*, in *British Journal of Anaesthesia*, Vol. 59, p. 1322.

DUNDEE, J.W., *et al.*, 1988, *P6 acupressure reduces morning sickness*, in *Journal of the Royal Society of Medicine*, Vol. 81, pp. 456–7.

EDSON, Lee, Fall 1991, *Newsletter of the National Association of Sciencewriters*, Vol. 39, no. 3.

EVANS, Christopher, 1973, *Cults of Unreason*, George G. Harrap, London.

EYSENCK, Hans J., and Sargent, Carl, 1982, *Explaining the Unexplained*, Weidenfeld & Nicolson, London.

EYSENCK, H.J., 1970, *The Structure of Human Personality*, Methuen, London.

EYSENCK, H. J., 1988, *Personality, stress and cancer: prediction and prophylaxis*, in *British Journal of Medical Psychology*, Vol. 61, pp. 57–75.

FESTINGER, Leon, 1962 edn., *A Theory of Cognitive Dissonance*, Stanford University Press, California.

FLEISCHMANN, M., and Pons, S., 1989, *Electrochemically induced nuclear fusion of deuterium*, in *Journal of electroanalytical chemistry*, Vol. 261, pp. 301–8, and Vol. 263, 187–8.

FOLEY, John, 1981, *The Boilerplate War*, W.H. Allen, London.

FOX, Edward, 1992, *Believe it or not*, in *The Independent Magazine*, November, pp. 45–52.

FRY, E.N.S., 1986, *Acupressure and postoperative vomiting*, in *Anaesthesia*, Vol. 41, pp. 661–2.

GAUQUELIN, M., 1969, *The Cosmic Clocks*, Peter Owen, London.

GHANTA, V.J., *et al.*, 1985, *Neural and environmental influences on neoplasia and conditioning of NK activity*, in *Journal of Immunology*, Vol. 135, pp. 848–52.

GOODCHILD, Peter, 1980, *J. Robert Oppenheimer*, BBC Books, London.

GURWITSCH, Alexander, 1937, *Mitogenic Analysis of the Excitation of the Nervous System*, North Holland, Amsterdam.

HAHN, Otto, and Strassman, Fritz, 1939, *Uber den Nachweis und das Verhalten der bei der Bestrahlung des Urans mittels Neutronen entstehended Erdalkalmetalle*, in *Die Naturwissenschaften*, Vol. 27, p. 15.

HAINES, A.P., *et al.*, 1987, *Phobic anxiety and ischaemic heart disease*, in *British Medical Journal*, Vol. 295, pp. 297–9.

HALL, Barry G., Sept. 1990, *Spontaneous point mutations that occur more often when advantageous than when neutral*, in *Genetics*, Vol. 126, pp. 5–16.

HALL, Robert N., (ed.), 1989, *Pathological science*, in *Physics Today*, October, pp. 36–48.

HANSEL, C.E.M., 1980, *ESP and Parapsychology: a Critical Re-evaluation*, Prometheus, Buffalo.

HASTED, John, 1981, *The Metal Benders*, Routledge & Kegan Paul, London.

HOLLAENDER, A., and Claus, W.D., 1935, *Journal of the Optical Society of America*, Vol. 25, p. 270.

HORROBIN, David, 1982, *In praise of non-experts*, in *New Scientist*, 24 June, pp. 842–4.

HUIZENGA, John, 1992, *Cold Fusion: Scientific Fiasco of the Century*, University of Rochester Press.

JUERGENS, Ralph, 1968, *Minds in chaos*, in *American Behavioural Scientist*, September.

KANT, Immanuel, 1969 edn., *Critique of Pure Reason*, J.M. Dent, London.

KELLY, Fred C., 1944, *The Wright Brothers*, Harrap, London.

KILNER, Walter J., 1975 edn., *The Human Atmosphere*, Samuel Weisner, New York.

KIRLIAN, Semyon D., and Kirlian, Valentina H., 1968, *Investigation of Biological Objects in High Frequency Electrical Fields etc.*, Alma Ata, USSR.

KOESTLER, Arthur, 1971 edn., *The Ghost in the Machine*, Pan, London.

KOESTLER, Arthur, 1974 edn., *The Case of the Midwife Toad*, Pan, London.

KUHN, Thomas S., 1970 edn., *The Structure of Scientific Revolutions*, University of Chicago Press.

KURTZ, Paul, 1976, *Committee to Scientifically Investigate Claims of the Paranormal and Other Phenomena*, in *The Humanist*, May/June, p. 28.

LAFLEUR, Laurence J., 1951, *Cranks and scientists*, in *The Scientific Monthly*, November, pp. 284–90.

LANGMUIR, Irving (See HALL, Robert N.).

LATIMER, W.M., and Young, H.A., 1933, *The isotopes of hydrogen by the magneto-optic method*, in *Physical Review*, Vol. 44, p. 690.

LAURIE, Peter, 1972, *Beneath the City Streets*, Penguin, London.

LEWIS, C.S., 1983 edn., *That Hideous Strength*, Pan, London.

LUCKHARDT, S.C., May 1992, *Technical appendix to D. Albagli, et al., Journal of Fusion Energy, 1990, Calorimetry Error Analysis*, MIT Report PFC/RR-92-7.

MACINTYRE, Donald, 1957, *Jutland*, Evans Brothers, London.

MAIER, S.F., and Seligman, M.E.P., 1976, *Learned helplessness: theory and evidence*, in *Journal of Experimental Psychology*, Vol. 105, pp. 2–46.

MALLOVE, Eugene, 1991, *Fire From Ice*, John Wiley, New York.

McCREERY, Charles, 1967, *Science, Philosophy and ESP*, Faber & Faber, London.

McGHEE, J.L., and Lawrentz, M., 1932, *Test for element 87 by the use of Allison's magneto-optic apparatus*, in *Journal of the American Chemistry Society*, Vol. 54, p. 405.

MEDAWAR, Sir Peter, 1985, *The Limits of Science*, Oxford University Press.

MELTON, J. Gordon, *et al.*, 1990, *Skeptics and the New Age*, in *New Age Encyclopaedia*, Gail Research, Detroit, Mich.

MILGRAM, S., 1974, *Obedience to Authority*, Tavistock, London.

MILLER, Henry W., 1930, *The Paris Gun*, George G. Harrap, London.

MILLER, Russell, 1992, *Resurrecting Lorenzo*, in *Sunday Times Magazine*, 20 December, pp. 38–44.

MILTON, Richard, 1993 edn., *The Facts of Life: Shattering the Myths of Darwinism*, Corgi, London.

MOSS, Thelma, 1981, *The Body Electric*, Paladin, London.

MOSS, Thelma, and Johnson, K., 1972, *Radiation Field Photography*, in *Physics*, July.

NASH, L.K., 1956, *The origin of Dalton's chemical atomic theory*, in *Isis*, Vol. 47, pp. 101–16.

NIXON, P.G.F., 1991, *Cardiovascular Health Promotion and Rehabilitation*, in BHMA response to the Government's Green Paper.

NORMAN, Bruce, 1984, *Here's Looking at You*, BBC, London.

OLDFIELD, Harry, and Coghill, Roger, 1988, *The Dark Side of the Brain*, Element, Dorset.

O'REGAN, B., 1987, *Healing, remission and miracle cures*, Special Report of the Institute of Noetic Sciences, Vol. 3, p. 14.

ORNISH, Dean, *et al.*, 1990, *The lifestyle heart trial*, in *Lancet*, Vol. 336, pp. 129–33.

PENFIELD, Wilder, 1975, *The Mystery of the Mind: a Critical Study of Consciousness and the Human Brain*, Princeton University Press.

PIAGET, Jean, 1967, *Language and Thought of the Child*, Routledge & Kegan Paul, London.

PICCARDI, Giorgio, 1962, *The Chemical Basis of Medical Climatology*, C.C. Thomas, Springfield, Ill.

PODMORE, Frank, 1909, *Mesmerism and Christian Science*, Methuen, London.

POLLEN, Anthony, 1980, *The Great Gunnery Scandal*, Collins, London.

POPPER, Karl, 1959, *The Logic of Scientific Discovery*, Unwin Hyman, London.

POPPER, Karl, 1963, *Conjectures and Refutations*, Routledge & Kegan Paul, London.

POPPER, Karl, 1977 edn., *The Open Society and its Enemies*, Routledge & Kegan Paul, London.

PRICE, George R., 1955, *Science and the supernatural*, in *Science*, Vol. 122, pp. 359–67.

PRICE, George R., 1972, *Apology to Rhine and Soal*, in *Science*, Vol. 175, p. 359.

PUTHOFF, Harold (See Targ, Russell).

RADIN, Dean I., and Nelson, Roger D., 1989, *Consciousness-related effects in random physical systems*, in *Foundations of Physics*, Vol. 19, pp. 1499–514.

RANDI, James, 1975 edn., *The Magic of Uri Geller*, Ballantine, New York.

REES, W.D., *et al.*, 1967, *Mortality of bereavement*, in *British Medical Journal*, Vol. 4, pp. 13–16.

REICH, Wilhelm, 1971, *Selected Writings*, Vision Press, London.

REICH, Wilhelm, 1976 edn., *Character Analysis*, Vision Press, London.

REICHENBACH, Karl von, 1850, *Researches on Magnetism, Electricity etc. in Relation to the Vital Force*, (trs. William Gregory), Taylor, Walton & Maberly, London.

ROKEACH, M., 1960, *The Open and Closed Mind*, Basic Books, New York.

SCHACTER, S., 1951, *Deviation, rejection and communication*, in the *Journal of Abnormal and Social Psychology*, Vol. 46, pp. 129–44.

SCHIFF, L.I., 1961, *A report on the NASA conference on experimental tests of relativity theories*, in *Physics Today*, Vol. 14, pp. 42–8.

SCHLEIFER, S.J., *et al.*, 1983, *Suppression of lymphocyte stimulation following bereavement*, in *Journal of the American Medical Association*, Vol. 250, pp. 374–7.

SHARPE, Robert, 1991, *The War Against Cancer*, in *Outrage*, the Journal of Animal Aid, Jun/Jul, no. 74.

SHELDRAKE, Rupert, 1988 edn., *A New Science of Life*, Paladin, London.

SIMONTON, O.C., *et al.*, 1980, *Getting Well Again*, Bantam, London.

SMITH, Richard, 1988, *Research begins at forty*, in *New Scientist*, 30 June, pp. 54–8.

SPIEGEL, David, *et al.*, 1989, *Effects of psychosocial treatment on survival of patients with metastatic breast cancer*, in *Lancet*, 14 October, pp. 888–91.

STRATTON, George M., 1897, *Vision without inversion of the retinal image*, in the *Psychological Review*, Vol. 4, pp. 341–60.

STURROCK, Peter, 1988, *Brave new heresies*, in *New Scientist*, 24/31 December, pp. 49–51.

SWARTZ, Mitchell R., 1992, *Reexamination of key cold fusion experiment by the MIT Plasma Fusion Laboratory*, in *Fusion Facts*, newsletter of Fusion Information Center, University of Utah, Vol. 4, No. 2, August, pp. 27–41.

TALBOT, Frederick A., 1912, *Steamship Conquest of the world*, William Heinemann, London.

TARG, Russell, and Puthoff, Harold, 1974, *Investigating the Paranormal*, in *Nature*, 18 October.

TAYLOR, Gordon Rattray (ed.), 1982, *The Inventions That Changed the World*, Reader's Digest, London.

TAYLOR, John, 1976, *Superminds*, Pan, London.

TAYLOR, John, 1980, *Science and the Supernatural*, Temple Smith, London.

THOMPSON, Sylvanus P., 1910, *The Life of Sir William Thomson, Baron Kelvin of Largs*, London.

THORNTON, E.M., 1983, *Freud and Cocaine*, Blond and Briggs, London. (Published in the US as *The Freudian Fallacy*, 1983, Doubleday, New York.)

TOMPKINS, Peter, and Bird, Christopher, 1974 edn., *The Secret Life of Plants*, Penguin, London.

TRENCH, Brinsley le Poer, 1974, *Secret of the Ages*, Souvenir Press, London.

VELIKOVSKY, Immanuel, 1973 edn., *Earth in Upheaval*, Sphere, London.

VELIKOVSKY, Immanuel, 1973 edn., *Worlds in Collision*, Sphere, London.

VON BRAUN, Wernher, 1971 edn., *Space Frontier*, London.

WATSON, James D., 1968, *The Double Helix*, Weidenfeld & Nicolson, London.

WICKRAMASEKERA, Ian E., 1988, *Clinical Behavioral Medicine*, Plenum Press, New York.

Other Sources

UK

The Times, 28 January 1926.
The Independent, 20 April 1989.
Daily Telegraph, 1 April 1989.
Daily Telegraph, 18 April 1989.
Daily Telegraph, 2 May 1989, p. 1.
Daily Telegraph, 15 May 1989, p. 12.
Daily Telegraph, 12 June 1989.
HMSO, 1991, *The Health of the Nation*. Government Green Paper on the future of the Health Service.
HMSO, 1993, *Government Review of R & D 1992*.
HMSO, 1993, *Britain 1993*.
Nature, 18 October (1974).
Nature, 293, 245, 246 (1981).
Nature, 344, 365 (1990).
New Scientist, 18 August 1988, p. 19.
New Scientist, 22 April 1989, p. 27.
New Scientist, 29 April 1989, pp. 22–3.
New Scientist, 6 May 1989, p. 26.
Science Fiction?, 1986, television film written and produced by Hilary Lawson and broadcast in the *Horizon* series by BBC2.
The Greenhouse Conspiracy, 1990, television film written and produced by Hilary Lawson and broadcast on Channel 4.

US

American Behavioural Scientist, September 1968.
Cancer Patient Survival: What Progress has Been Made?, US General Accounting Office, 1987.
Cold Fusion 1992: Basic Facts (newsletter published by Clustron Sciences Corporation, Vienna, Va.).
The Independent, 22 October 1903.
Missiles and Rockets, 18 January 1965.
New York Times, 11 June 1950.
Scientific American, 13 January 1906.
Time, 4 November 1991, p. 75.

Subject Index

People Index